MODULAR SCIENCE

BOOK 1

MODULE 1 *Environments*

MODULE 2 *Pollution*

MODULE 3 *Humans as organisms*

Writing team: Mike Hiscock, Will Deloughry, Phil Naylor

This book is about science. How does science affect you, and the world around you?

The three modules in this book each tell you how science affects some aspect of your life. Some of the questions and activities in the modules are there to help you understand what you have read. Others help you think scientifically, and handle scientific data – facts or information which may be words or numbers.

Turn the page to find out more about the modules in this book.

Heinemann Educational Books

MODULE 1 Environments

*Any place where plants or animals are able to live is called an **environment**. This module looks at several different environments and how they support the plants and animals that live there.*

1.1	What are environments	**2**
1.2	A woodland community	**4**
1.3	Who's eating who?	**6**
1.4	A changing environment	**8**
1.5	The energy factory	**10**
1.6	Using and losing energy	**12**
1.7	Removing nature's waste	**14**
1.8	Chemical merry-go-round	**16**
1.9	Life in the balance	**18**
1.10	Counting up the numbers	**20**
1.11	Hedgerows – an ideal home?	**22**
1.12	The disappearing otter	**24**
1.13	Swan death – who's to blame?	**26**
1.14	Destroying natural habitats	**28**
1.15	Decisions affecting the environment	**30**

MODULE 2 Pollution

*Do you think there is a link between watching the TV and releasing poisonous gases into the air? Many aspects of everyday life can damage the environment – this type of damage is called **pollution**. This module looks at pollution and how to avoid it.*

2.1	What is pollution?	**32**
2.2	Take care of the air	**34**
2.3	Acid rain, deadly downpour	**36**
2.4	Nature's warning signs	**38**
2.5	Mucking about with water	**40**
2.6	Murky waters	**42**
2.7	Protecting life in the sea	**44**
2.8	Manufacturing pollution	**46**
2.9	Radioactive waste	**48**
2.10	Pesticides – a growing problem	**50**
2.11	Polluting your body	**52**
2.12	Noise pollution	**54**
2.13	What a waste!	**56**
2.14	Products that don't pollute	**58**
2.15	The problems with pollution	**60**

MODULE 3 *Humans as organisms*

All plants and animals are organisms. Many organisms do not have much control over the conditions in which they live. As a human being, you are unusual because you can make many choices about how you live. To help you make the right choices, this module looks at how the human body works.

3.1 Healthy organisms **62**

3.2 Using energy **64**

3.3 What are you eating? **66**

3.4 Eating the right amount **68**

3.5 What happens when you eat? **70**

3.6 Every breath you take **72**

3.7 Blood – supplying your needs **74**

3.8 Your heart – a muscular pump **76**

3.9 Waste – life's leftovers **78**

3.10 Muscle & bone, a joint effort **80**

3.11 Getting hurt **82**

3.12 Controlling your reactions **84**

3.13 Measuring fitness **86**

3.14 Keeping fit **88**

3.15 What a life you lead! **90**

Index **92**

Acknowledgements **94**

1.1 What are environments?

When you look out of your window what do you see – fields, a garden or a street with other houses? What you see is the environment in which you live. The plants around you, the houses, other people, the animals, the air you breathe and the ground you walk on – these are all part of your environment.

In this module you will be trying to find out answers to the following questions about environments. There are many other questions too!

Why do some environments have lots of types of wildlife?

A simple hedgerow may contain . . .

. . . up to 30 types of shrubs and trees

. . . up to 65 types of birds which nest in hedges.

. . . up to 1500 types of insects.

. . . up to 600 types of wild flowers.

. . . up to 20 types of small mammals – voles, mice, hedgehogs and so on.

Why do some environments have only a few types?

If you look around a busy town centre you will not see many types of birds. The birds shown are usually common sights. Why are there so many of these and only very few other types?

What effect do we have on our environment?

These crops are being sprayed with insecticide - a poison that stops them being eaten by insects.

How do you think the insecticide will affect the life in the hedgerows around the field?

This digger has been used to remove a hedgerow – now there will be only one big field instead of two small ones. The uprooting of hedgerows to make bigger fields has been going on for many years. This is to let large machinery, like combine-harvesters, move around more easily. What effect will this have on farmland wildlife?

What can we do to have a good effect?

The countryside on the edge of towns could be used for our enjoyment – but it is frequently neglected. These people are clearing away rubbish that has been dumped into a pond. Do you think all their hard work is worth it?

There are many neglected sites in towns and cities – such as old churchyards, disused railway lines and overgrown wasteland. Planting wild flowers and trees can improve these environments by attracting insects, birds and other animals. Whose responsibility is it to do this?

Environments are like life-support systems for all the animals and plants that live in them – including you! To find out more, read on.

1.2 A woodland community

Hidden life

Have you ever walked through a wood and noticed how few animals there are – maybe just a few common birds and insects? Don't be misled. A wood is full to the brim with a wide variety of wildlife.

See how many examples of woodland wildlife you can find on this picture.

Tawny Owls hunt voles and mice at night. They nest in holes in trees.

Grey squirrels are common woodland animals. They damage trees by feeding on young shoots and leaf buds. In the autumn they bury acorns which they will eat during winter.

Badgers live in burrows called sets. They hide from predators in thick woodland undergrowth, and feed on worms, insects and many types of roots and berries.

Oaks and other large trees provide food and shelter for mammals birds and hundreds of insects.

Dog's Mercury and other wild flowers grow well in the shade.

Fallen leaves are broken down by fungi on the woodland floor.

Habitat sweet habitat

The place where an animal or plant lives is called its **habitat**. The wood is the habitat of all the animals and plants shown in the picture above – it is their home. Each animal needs food, shelter, and a place to hide from predators. The plants need light, water and nutrients.

Woodland animals can find many types of food and many different places to shelter and hide. These are provided by the wide variety of plants in the wood – the tall trees, bushes and flowers on the ground.

Other habitats

The wood is only one example of a habitat. Other examples are a pond, a river, and a sand dune. Each of these is a different habitat. This picture shows how very different sand dunes are from woods.

Marram grass is the most common sand dune plant – it binds sand together and stops it being blown about by winds.

Woodland plants need rich soil to grow in but the dune plants are able to grow in sand. You can see that the main plants in the dunes are grasses. There are no large trees. Do you think that woodland animals would be able to survive in the dune habitat?

Community life

All the wildlife that share a habitat form a community – they depend on each other somehow. For example, the hundreds of plants and animals in a wood need food; where does it come from?

Food makers . . .

Plants are the most important members of the woodland community, particularly the trees. Plants can capture the energy in sunlight and use it to make food. Details are given on page 10. As plants make their own food we call them **producers**.

Plants use some of the food they make to grow. They store the rest in food stores such as acorns, berries and nuts. Other members of the woodland community use plants as a food supply.

*These tortoiseshell caterpillars (**herbivores**) are chewing through nettle leaves (**producers**).*

. . . and food takers.

All the animals in the community are **consumers**. They eat plants or other animals as a source of food. There are three different types of consumers.

Some animals, such as caterpillars, consume plant material. These are the **herbivores**. They are often very choosy about what they eat, feeding on the leaves, bark, sap, nectar, or seeds of particular plants.

Other animals eat the herbivores. These are the **carnivores**. They may in turn be eaten by other animals – the **top carnivores**.

Many animals and plants use the dead remains of other living things as food – worms and fungi are examples. We call them **decomposers** – they break down (decompose) the things they feed on into wastes. Producers can use these wastes as raw materials to make food.

*Butterflies (**herbivores**) drink the nectar found in flowers like this thistle.*

1 Give examples of woodland wildlife that are:
- **a** Producers
- **b** Herbivores
- **c** Carnivores
- **d** Decomposers

2 In what ways does the woodland provide food and shelter for the animals shown in the picture on the opposite page?

3 Carnivores are not plant-eaters, but the woodland carnivores depend on plants for their food. Explain why.

4 **Omni** means *all kinds of*. So what do you think an **omnivore** is? Can you give an example of one?

5 Make a list of plant materials that herbivores use as food. Some examples are shown on this page. You should be able to think of many others.

1.3 Who's eating who?

Food chains

Even though some animals may never eat any plants, they still depend on the plants. But how can a carnivore like a stoat depend on a primrose? This picture shows you how.

Each plant or animal is a link in a **food chain**. In every food chain the first link is a producer – a plant that can make its own food by photosynthesis. The second link is a herbivore and the third is a carnivore. There may be more carnivores in the chain too. Here is an example:

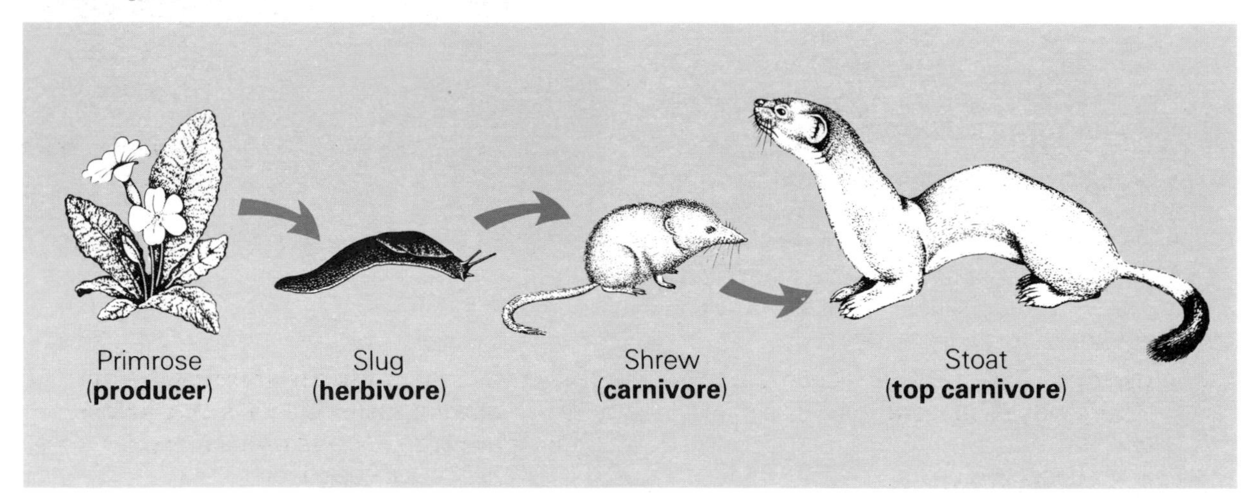

1 The food chain in the first picture can be shown more simply like this:

primrose → rabbit → stoat

See if you can write down other examples of food chains. Some animals which you could use are given in the box:

A varied diet

In the food chains on the opposite page stoats are shown eating shrews and rabbits. Shrews will eat a variety of animals found among the rotting leaves on the woodland floor – worms, slugs and beetles, as well as woodlice. Some animals are not at all choosy about what they eat – badgers eat hedgehogs, mice, beetles and worms. They will even visit gardens and eat scraps left out at night!

Some individual food chains do not tell us very much about an animal's eating habits. To understand more about the woodland community we need to look at other ways of showing who eats who.

Food webs

If you have tried Question 1 you may have found that some of your food chains could be connected together. The diagram below shows the feeding links between a small number of woodland animals and plants. A group of food chains connected together in this way is called a **food web**. Food webs tell you more than food chains. A food web can give you an overall view of how the members of a community are linked together.

Look carefully at the picture and see who is eating who.

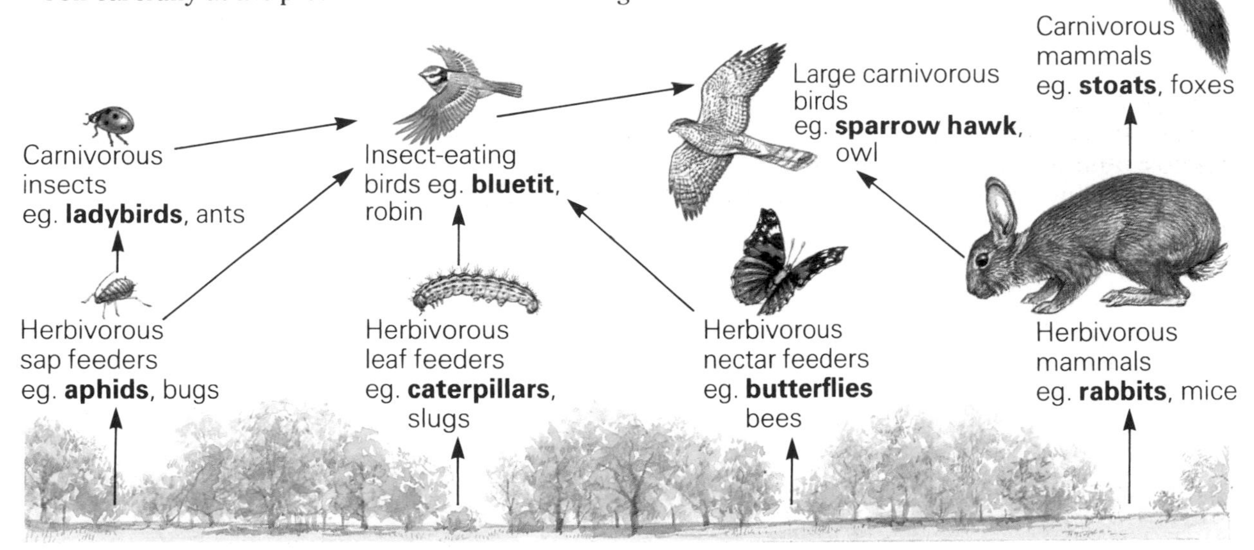

2 Use the food web shown above to give an example of a herbivore, a carnivore and a top carnivore.

3 Name an insect that
a eats other insects
b feeds on nectar
c eats the leaves of trees.

4 Use the food web above to produce two food chains, one involving three organisms and the other four organisms.

5 Even a large wood has only a few large carnivorous birds like the sparrowhawk. Why is this so? What would happen if there were too many of these birds?

6 What would happen to the animals shown in the food web if a large number of the trees in the wood were chopped down or died?

7 How will winter affect the animals and plants shown in the food web? Explain your answer carefully.

1.4 A changing environment

A gardener's nightmare!

Imagine what would happen if a large garden was left to grow wild after being completely dug over. The pictures below show the changes that could take place. Can you see what happens as the garden gets more and more overgrown?

After 5 years

Goldfinches feed on small seeds – like those of the thistle

The garden becomes covered in grasses and weeds.

After 20 years

Young trees and bushes provide food and nesting places for birds.

After 100 years

Once the trees are fully grown, there's more variety of food and so there's more variety of wildlife.

The first colonisers

The changes that take place in the garden happen very slowly. Some of the first plants to appear are those that grow from lightweight seeds, blown in by the wind. These are usually quick-growing plants. This means they can take advantage of the unoccupied garden. Thistles are an example of this type of plant.

Other plants follow and soon there is a wide variety of weeds. The weeds compete with each other for space to grow and for light and soil nutrients.

Thistle seeds are blown over a wide area by the wind. Can you see the seeds in the middle of each fluffy ball?

From weeds to trees

After a few years, young trees and shrubs grow up among the weeds. These are woody plants that get bigger each year. They take up more space and more nutrients from the soil. They also cast a shadow over the smaller plants. This **competition** from larger plants prevents the smaller plants from growing and they gradually die. Eventually the young trees get very tall and the garden becomes like a woodland.

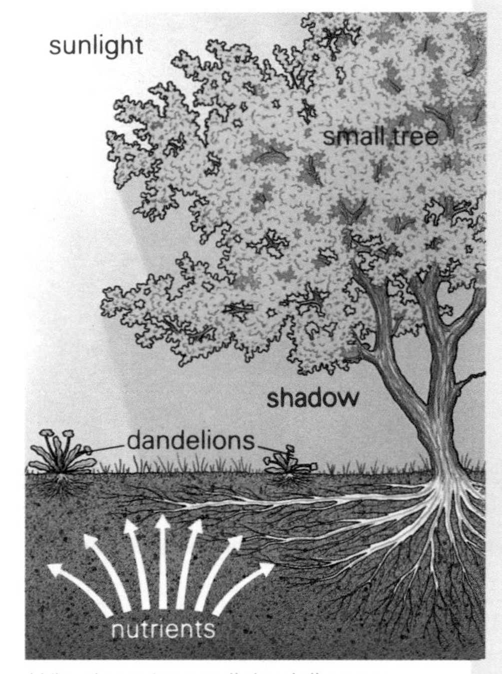

Why does the small dandelion get less sunlight, less space to grow and less nutrients from the soil?

The changing birdlife

The weed covered garden provides soil insects and worms for ground-feeding birds like starlings and blackbirds. Thistles and other plants provide food for seed-eating birds like goldfinches.

When the garden becomes covered with young trees and shrubs it provides even more shelter and nesting sites. Larger types of seeds, fruit and insects can now feed a wider variety of birds such as thrushes, chaffinches and bluetits.

Many more birds are able to live in the garden once it is full of large trees. Even the carnivorous owl can find enough food to satisfy its appetite.

Sudden death

The woodland garden would be changed dramatically if you chopped down the trees – no more food and cover for hundreds of animals. The trees take years and years to grow but it only takes a few minutes to chop them down. The habitat that has taken years to develop could be ruined in a single day.

1. Name an example of a ground-feeding bird, a seed-eating bird and a berry-eating bird.
2. Explain why many of the weeds in the garden disappear as the trees and shrubs became common plants in the garden.
3. Why are more types of birds found when there are trees in the garden instead of just weeds?
4. None of the plants that appear in the garden were planted. Try to find out the different ways that the plants could get there.

1.5 The energy factory

Plants – the food producers

Plants produce their own food. This is why they are called producers. All animals, including you, depend on plants for food. The method plants use to produce food is called **photosynthesis**. This picture shows you what happens.

Sunlight provides energy

Carbon dioxide gas is absorbed from the air

Sugars are made in the leaf – these are the plant's food

Oxygen gas is produced and released into the air

Water from the roots enters the leaf

1 Study the picture carefully. Pick out the materials the plant needs for photosynthesis. Which materials are **produced** by the process?

Making sugars, while the sun shines

The picture above shows that plants can produce the food they need from **carbon dioxide** and **water**. These chemicals are used to make **sugars**, which are energy-rich food substances.

Oxygen is also produced as a by-product. As plants photosynthesise they remove carbon dioxide from the air and add oxygen back to the air.

Storing the sun's energy

It takes energy to make photosynthesis happen. Plants use **chlorophyll**, a green chemical in their leaves, to absorb the energy in the sun's light. They 'lock up' this energy in sugars, ready to be released when it is needed. How does all this happen inside the leaf? This diagram tells you more.

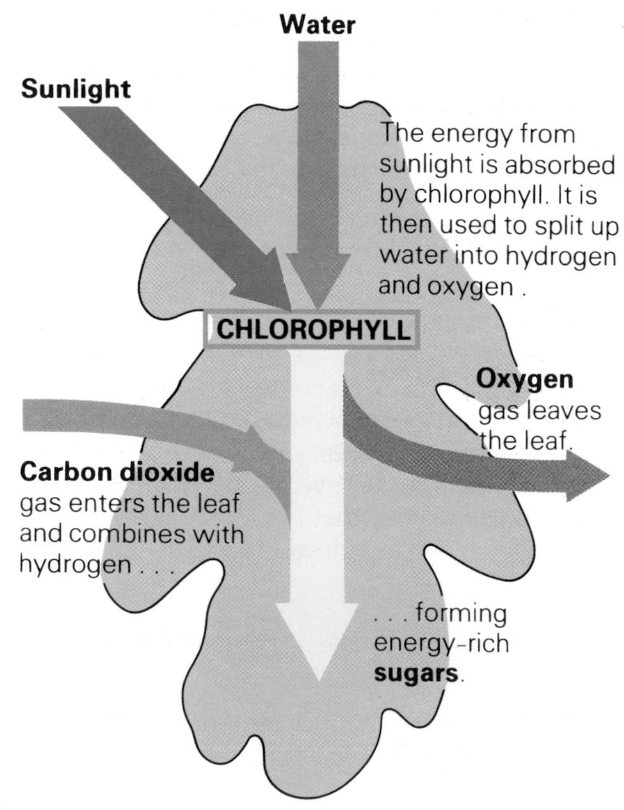

Photosynthesis – storing energy in sugars.

What happens to the sugars?

Photosynthesis can only happen in the parts of the plants that contain chlorophyll. Most chlorophyll is in the leaves, but the sugars made here can be transported to all parts of the plant.

The energy stored in these sugars can be released straight away if necessary. If the energy is not needed immediately, the sugars can be converted to other substances such as starch and cellulose. This can help the plant to keep some energy in reserve.

Sugars are simple food substances which are soluble in water. This makes it easy to transport them around the plant.

Starch is a complex substance which is not soluble in water. Storing food in an insoluble form means that much more can be packed into the plant. The plant can turn starch back into simple sugars for the release of energy.

Cellulose is another complex substance made from sugars. It helps to make plants stiff and tough. It is not soluble in water. The plant can *not* turn cellulose back into simple sugars – so it can not be used as a store of food.

Sugars, starch and cellulose are all **carbohydrates** – chemicals which are made from carbon, hydrogen and oxygen. The carbon and oxygen in carbohydrates come from the carbon dioxide in the air. The hydrogen in the carbohydrates comes from water inside the plant.

Grape vines absorb energy from sunlight and lock it up in sugars. The longer they stay in the sun, the sweeter they get!

Releasing the stored energy

Plants need energy just like you do. They use it to make the materials they need to grow. They use energy to help take in nutrients from the soil and to transport chemicals around the plant. They use a process called **respiration** to release energy locked up in sugars. Respiration releases energy stored during photosynthesis. This diagram tells you more.

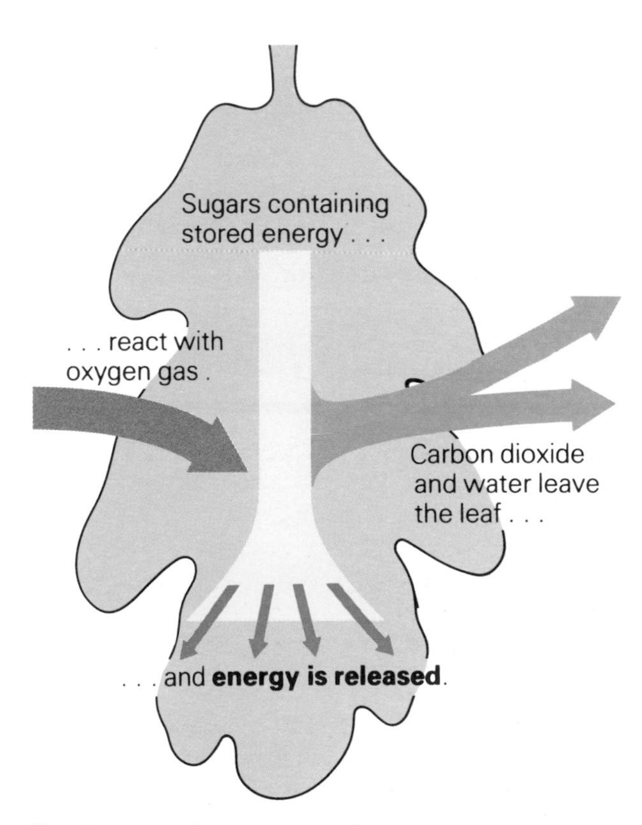

Respiration — releasing energy from sugars.

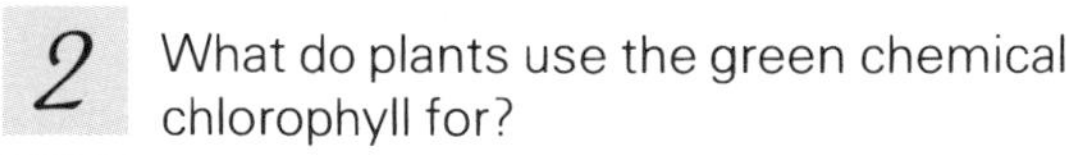

2 What do plants use the green chemical chlorophyll for?

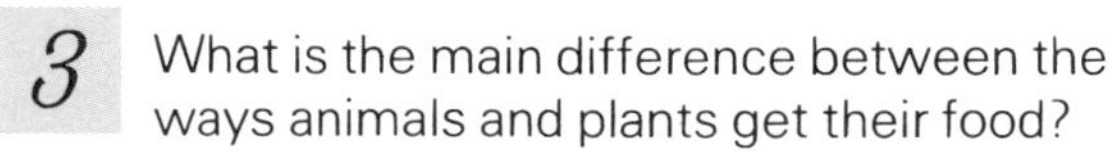

3 What is the main difference between the ways animals and plants get their food?

4 Make up a table to show what is used up and what is produced during respiration and photosynthesis.

5 Write a short paragraph explaining *in your own words* what photosynthesis is, and how it works.

1.6 Using and losing energy

Food chains and energy

All living things need energy from the sun to stay alive. Look at this food chain.

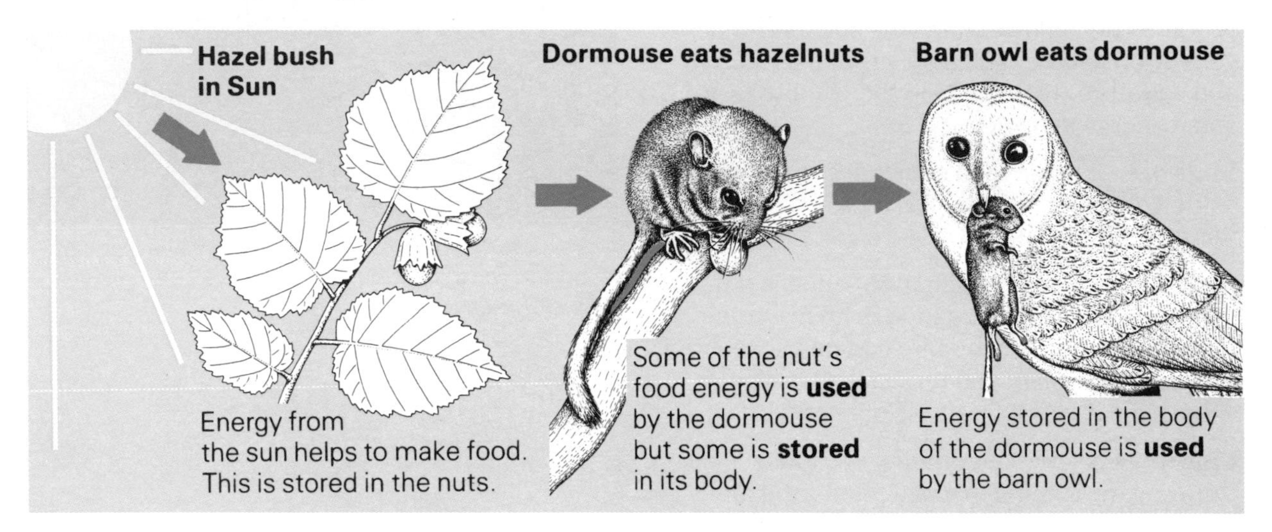

Each animal or plant in a food chain releases **energy** from its food by respiration. (You can find out more about respiration on page 11).

What is this energy used for?

Using energy . . .

All plants and animals, including you, need energy to do things. This diagram shows how a dormouse uses the energy from its food.

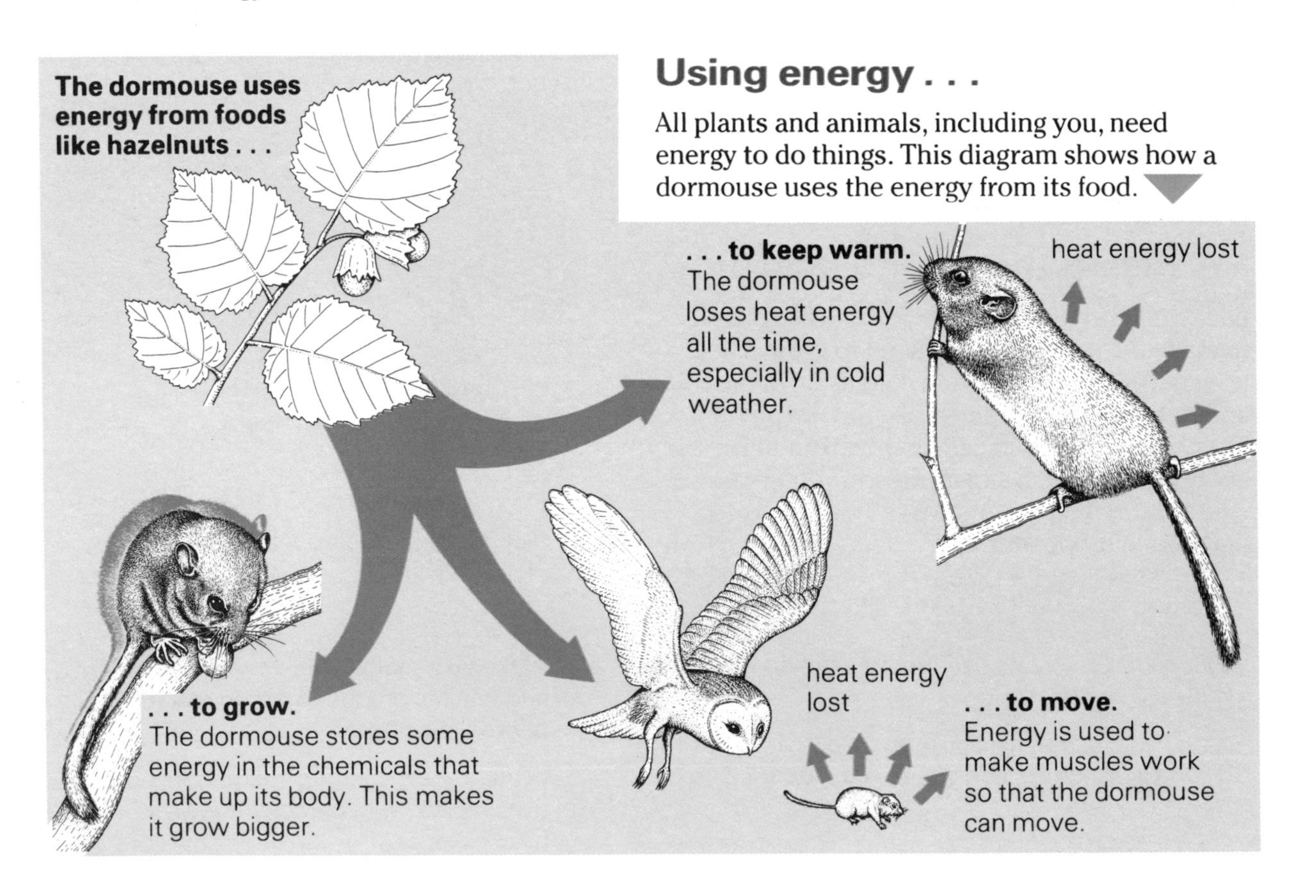

. . . and losing energy

Why do barn owls eat dormice? The owls do it to get the energy stored in the chemicals in the dormouse's body. But the energy that the dormouse has used to keep warm and to move has been lost from its body. This energy is **lost** from the food chain.

Not all the food eaten by the dormouse can be used. Some of it passes through the animal's gut and out as waste material – its 'droppings'. Some of the energy still stored in the waste material can be released by insects which feed on the droppings. Very small organisms called fungi and bacteria also feed on the waste.

This diagram shows how little of the energy in a dormouse's food is passed along the food chain.

Heat energy is **lost** as the dormouse moves and keeps warm.

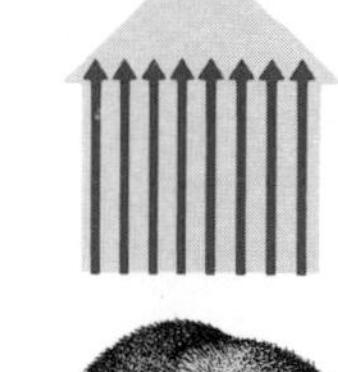

Food eaten contains energy stored in chemicals.

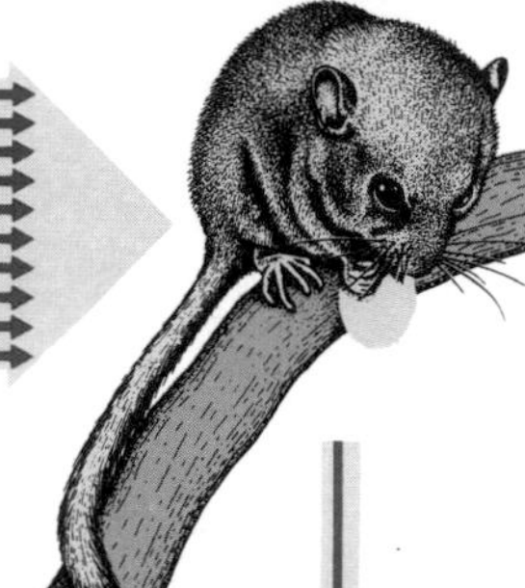

Energy is passed on to the animal that eats the dormouse.

Energy is **lost** in waste materials

Key
Each arrow represents $\frac{1}{10}$ of the energy taken in by the dormouse.

Energy flow

At each stage in a food chain some energy is stored in plants or animals. The rest is lost from the chain.

The hazel bush locks up a lot of energy into foods during photosynthesis but only a little of that energy will be passed on to the owl. This diagram shows where the energy is lost on its way along the food chain.

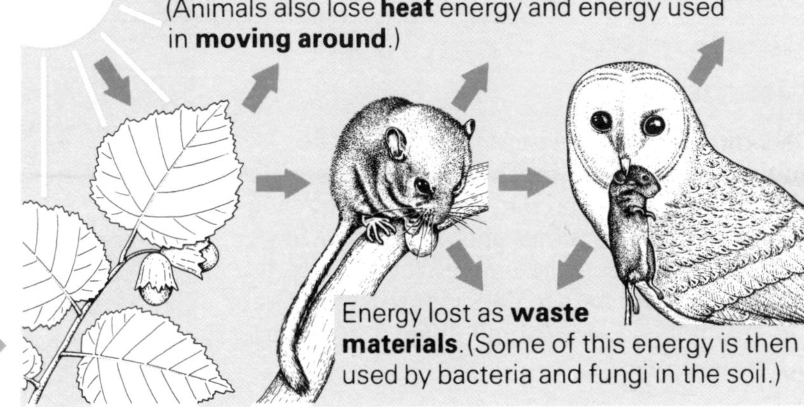

1 Copy and complete the following food chain.

? → dormouse → ?

2 Find out what fraction of the energy in the dormouse's food is:
a used to keep it warm and to move;
b lost in waste materials;
c passed along the food chain.

3 Here are two food chains:

wheat → human
grass → lamb → human

In which chain is the least energy lost? Explain your answer.

4 Some farmers produce lambs and other animals.
a why don't all farmers grow plant crops?
b why would it be a good idea for more farmers to grow crops?

5 You are the last link in many food chains. Most of these chains are short – there are only two or three living things in them. Why is this? See if you can make a food chain ending with you that involves four or more living things.

1.7 Removing nature's waste

Woodland waste

In autumn the ground is often covered with fallen leaves. Soon all these leaves disappear. Where do they go to? Do they just get blown away?

They are removed by nature's waste removers – animals and plants which live in the soil. Without them woods and parks would be littered with fallen leaves!

How are all these leaves cleared up?

Life on earth

When you first look at the soil what do you see? Just a collection of dirt particles and bits of plants. But look harder and you should find a large number of animals including worms, woodlice, beetles, slugs and snails. These animals live on the ground and among the soil particles. Many of them are **decomposers** – they use animal droppings, dead animals and dead plants as a source of food.

Decomposers – nature's bin men

Decomposers play an important part in the community they live in. They are nature's rubbish removers and they work fast, removing leaves, animal droppings and corpses in a few weeks. Not all waste is removed this quickly. It may take many years to decompose a tree. These pictures tell you more about different types of decomposers.

Earthworms pass soil through their bodies. They use the dead plant matter in the soil as their food.

'Burying beetles' hide dead animals below the soil and then lay their eggs in the decaying flesh.

Not all decomposers are animals. These fungi can completely decompose dead wood.

Recycling the rubbish

As well as tidying up the woodland floor, decomposers have another important role. They break down dead and waste materials into simple chemicals and release them into the soil. Some of these chemicals, such as nitrates and phosphates, are nutrients which plants need for growth. Plants can take them up through their roots. These chemicals were once parts of other living things but now can be re-used. This is called **recycling**.

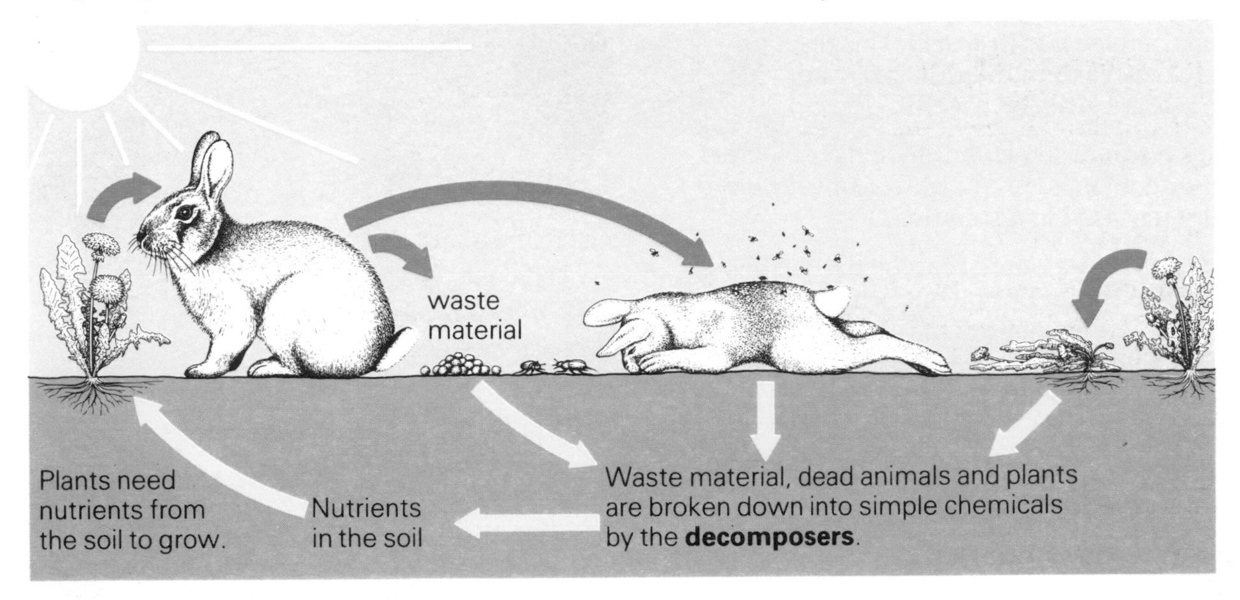

Disturbing the natural cycle

Not many of the plants grown by us are allowed to die and rot in the soil. Farmers and gardeners harvest their crops as soon as they ripen. As a result recycling cannot occur and the soil becomes less fertile – it doesn't contain enough of the nutrients plants need for healthy growth.

This problem can be solved by adding natural or artificial **fertilisers** to the soil to replace the lost chemicals. Gardeners use natural fertilisers. Their compost heaps contain rotting material that can be dug into the soil to add valuable nutrients. Composting garden waste is a way of recycling natural nurients back into the soil. Compost also makes the soil easier to dig. Manure can be used in the same way.

Most farm soils do not contain enough essential nutrients, so artificial fertilisers are added to improve food production.

1. What is a decomposer? Give three examples.
2. What useful things do decomposers do in an environment?
3. Explain how nutrients taken up from the soil by green plants can be used over and over again (recycled).
4. What methods can be used to improve the fertility of soil?
5. Some fungi can completely decompose wood. What effect will these fungi have on a wooden fence? How could you protect the fence?

1.8 Chemical merry-go-round

Releasing carbon dioxide

Plants and animals store energy in sugars and other carbohydrates in their bodies. When energy is needed it is released by **respiration**. To do this animals and plants **take in oxygen** and combine it with sugars. This reaction **releases carbon dioxide** and energy. More details are given on page 10.

This diagram shows the amounts of different gases contained in a lungful of air before and after it is used in respiration.

Imagine the amount of carbon dioxide being added to the air by millions of animals and plants. One estimate of the amount is 5 000 000 000 000 tonnes every year! But why is there always plenty of oxygen in the air?

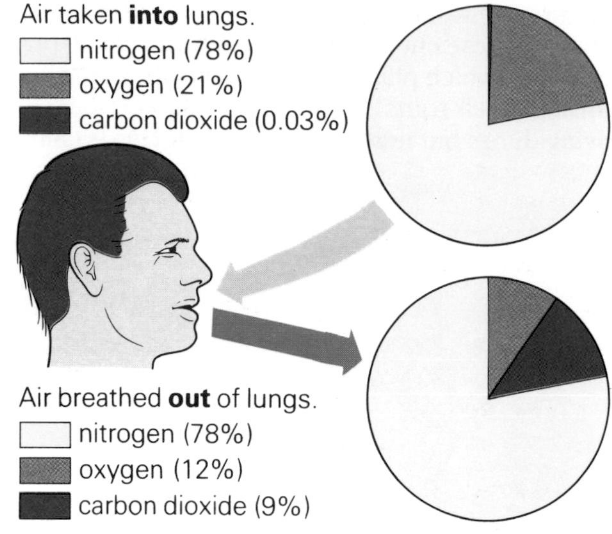

What do the pie charts tell you about how air is changed inside a person's lungs?

Using carbon dioxide

Plants produce their food by **photosynthesis** – the opposite of respiration. They **take in carbon dioxide**, water and energy. Sugars are made and the plants release oxygen.

This means that carbon dioxide and oxygen are involved in both respiration and photosynthesis. Look at this diagram.

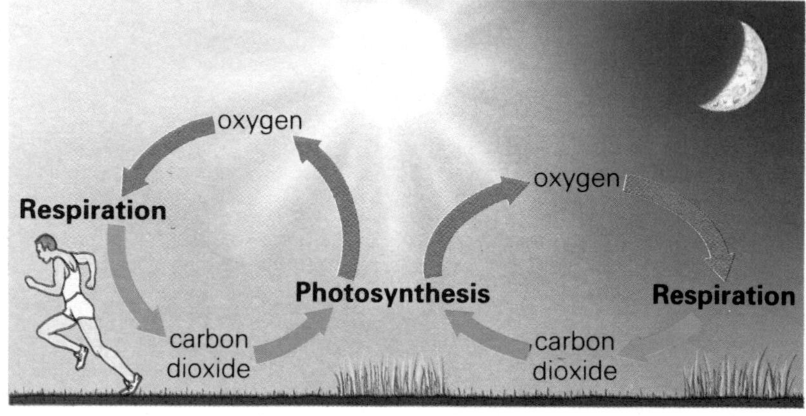

The carbon dioxide you breathe out may even be taken in by the grass you play on!

This piece of coal contains a fossil of a fern.

Other sources of carbon dioxide

When animals and plants die they stop respiring. The dead remains are used as food by **decomposers** – bacteria, fungi and animals that live in the soil. As decomposers break down dead remains, they release carbon dioxide into the air.

Decomposers can't always get at the dead remains. Fossil fuels like coil, oil and gas are the remains of plants which lived millions of years ago. The sugars (and other compounds) in these plants contain carbon. Instead of being decomposed they have been crushed under layers of mud and rock. The carbon from the dead remains stays locked up in the fuel.

On burning, carbon (from the fuels) reacts with oxygen (from the air) and carbon dioxide is produced. This process is called **combustion**.

Carbon goes round in circles

Industrial pollution needs to be carefully controlled to prevent too much carbon dioxide being formed.

Carbon goes round and round in the environment. Other chemicals go round and round too but carbon is a very important chemical. Carbon dioxide is a chemical which contains carbon. Plants use carbon dioxide to make sugars in photosynthesis. Animals eat plants to get the sugars. The animals use the sugars in respiration and produce carbon dioxide again. This means that the carbon in carbon dioxide can be used and then re-used – it is **re-cycled**.

There is a balance between respiration and photosynthesis which keeps the amount of carbon dioxide in the air at a steady level. But the huge amounts of fossil fuels burnt each day are interfering with this balance.

The diagram below shows how carbon moves from place to place in the environment. Because it goes round and round, this is called the **carbon cycle**. Study the diagram of this cycle. Spot the processes which add carbon dioxide to the air and the processes which remove it.

The carbon cycle.

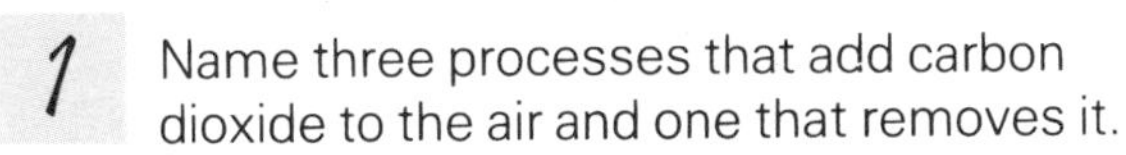

1 Name three processes that add carbon dioxide to the air and one that removes it.

2 Use the pie charts at the top of page 16 to find out the difference in per cent:

a between the oxygen in the air breathed in and the air breathed out;

b between the carbon dioxide in the air breathed in and the air breathed out.

What causes these differences?

3 While you are reading this you are producing about 250 cm^3 of carbon dioxide every minute. How much will your class produce in the next hour? How much will the whole school produce in this time?

4 Draw a diagram to show how oxygen in the environment is recycled. Use the diagram above as a guide: remember that oxygen and carbon dioxide move in opposite directions through the environment.

1.9 Life in the balance?

Looking below the waterline

Animals in an environment depend on plants for their food. They are linked together in a food web. An example of a food web is given on page 7.

Animals and plants that live in water are linked together in food webs too. The picture below shows life in an aquarium containing pond animals and plants. Look at the life in this sealed aquarium.

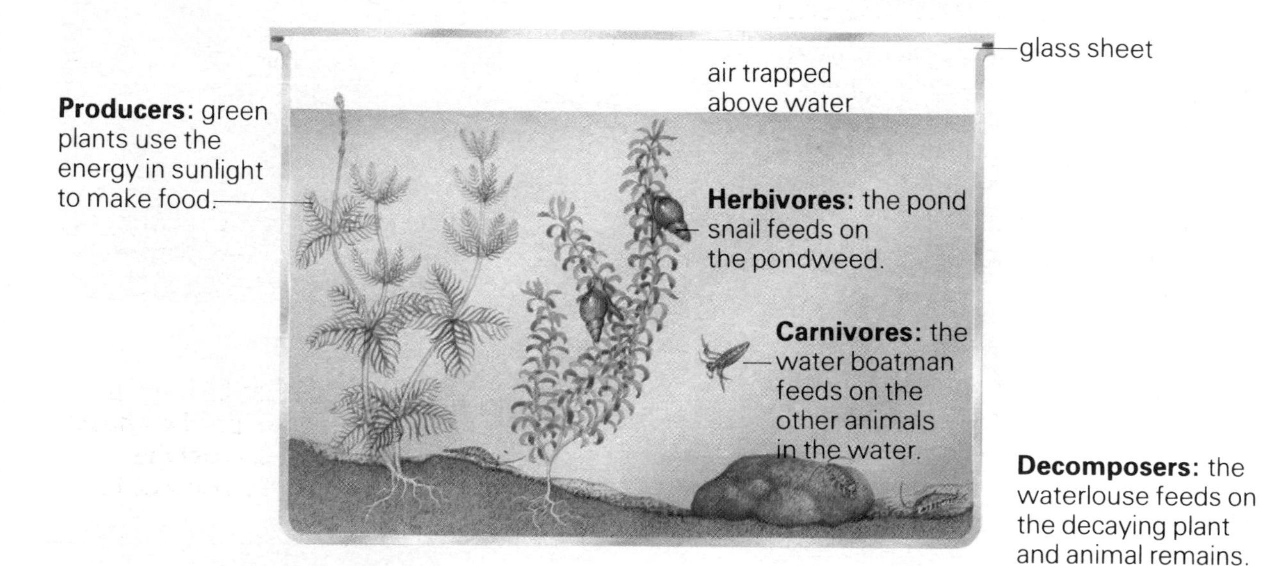

The waterlouse – one of the decomposers who help to keep the aquarium 'neat and tidy' and recycle nutrients too!

Being self-contained

The plants and animals in the aquarium will continue to survive even though the tank is sealed with a glass top and nothing is added to the water. All that is needed to keep the plants and animals alive is sunlight. Everything else that the plants and animals need comes from the living community in the aquarium.

Tidying up the waste

When the plants and animals in the aquarium die, they sink to the mud at the bottom. Their remains are used as food by **decomposers** – bacteria, fungi and some animals such as the waterlouse. They break down this waste into chemicals such as nitrates and phosphates. These chemicals are nutrients which the pond plants need to grow. The recycling of nutrients by decomposers means that there is a regular supply for the plants in the aquarium.

Recycling oxygen and carbon dioxide

All living things release the energy they need from sugars by **respiration**. They need oxygen to do this. Pond animals and plants use oxygen dissolved in the pondwater. As they respire they release carbon dioxide into the water.

Plants need carbon dioxide to make food by **photosynthesis**. As they do this they remove carbon dioxide from the water and add oxygen. This is the *opposite* of the process of respiration.

The **balance** between respiration and photosynthesis in the water keeps oxygen and carbon dioxide at fairly steady levels. The amounts of these chemicals will change only slightly during each day. If the number of animals in the aquarium was increased the balance would be upset. The amount of oxygen in the water would drop and the amount of carbon dioxide would increase.

Bubbles of oxygen released by the pondweed during photosynthesis. Look at how much oxygen has collected at the top.

Testing for carbon dioxide

You can investigate changes in the amount of carbon dioxide in pondwater by adding a harmless chemical called an **indicator** to it. When the amount of carbon dioxide in the water is low, the indicator is blue. When the amount of carbon dioxide is high, the indicator becomes green. This diagram shows how the colour of the indicator gives you an idea of what is going on inside the tube.

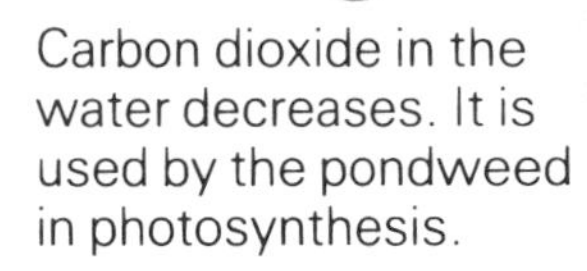

Carbon dioxide in the water decreases. It is used by the pondweed in photosynthesis.

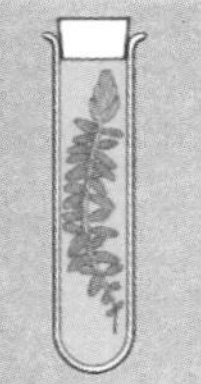

Carbon dioxide in the water increases as the pondweed is only respiring.

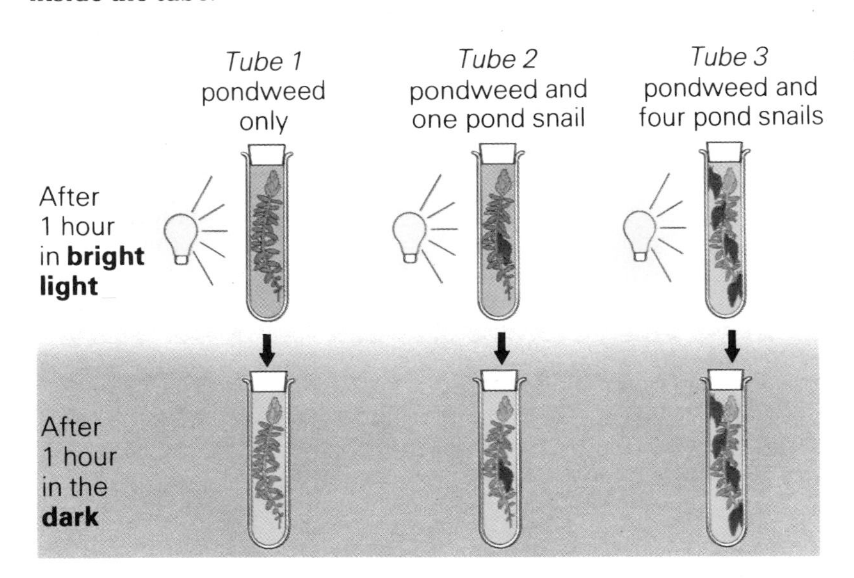

This method can be used to investigate how pondweed and pond snails can affect the amount of carbon dioxide in pondwater.

Look at these diagrams carefully and use them to answer Questions 3, 4 and 5 below.

1 What is needed from outside of a sealed aquarium to keep a community alive inside?

2 Where do the nutrients in the aquarium come from?

3 What processes occur in *tube 1* (see above)
- **a** in the light?
- **b** in the dark?

4 What processes occur in *tube 2* (see above)
- **a** in the light?
- **b** in the dark?

5
- **a** Why is *tube 3* the only one in which the indicator remains green in the light?
- **b** What will happen to the pond snails in *tube 3* if it is left sealed? Explain your answer.

1.10 Counting up the numbers

Find out what is there

How would you find out what animals and plants live on your school field? How would you know how many of each are present? You could not go out and count every single plant and animal you find. That would take you far too long – you will need to **sample** the area. Then you only need to identify and count the animals and plants in your sample.

Area A No weed killer used

Plants don't run away . . .

It is easy to sample an area to find out what plants grow there. One common way of sampling is to use a **quadrat**. These diagrams show you what a quadrat is and how to use it.

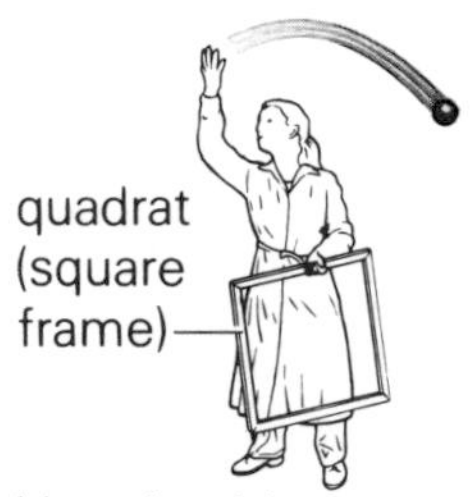

You should not choose where to place a quadrat. Throw a ball over your shoulder and place the quadrat where the ball lands.

grass: *several* (covers over $\frac{3}{4}$ of the area)

clover: *1* (covers much less than $\frac{1}{4}$ of the area)

daisies: *4* (covers less than $\frac{1}{4}$ of the area)

dandelion: *1* (covers less than $\frac{1}{4}$ of the area)

First **identify** the plants inside the quadrat. Next **count** the number of each type of plant. Then estimate the **area covered** by all.

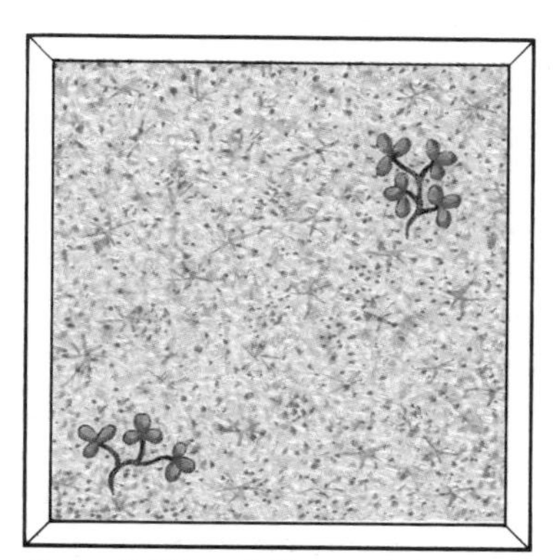

Area B Weedkiller used

1
a What area does each type of plant cover in quadrat A and quadrat B?
b Which type of plants have been affected by the weedkiller?

. . . but animals do

Animals are more difficult to sample because they either hide or run away. All the **different types** of animals living in the same environment make up the animal **community**. To find out what animals there are in an environment you must first catch them. These diagrams show how you can catch animals from different environments.

Ways of catching animals.

Comparing areas

Each of the collecting methods can be used to compare animals in different environments. Pitfall traps, for example, can be used to compare the animals found in a field and in a wood. The animals in the traps should be identified, counted and recorded in a chart.

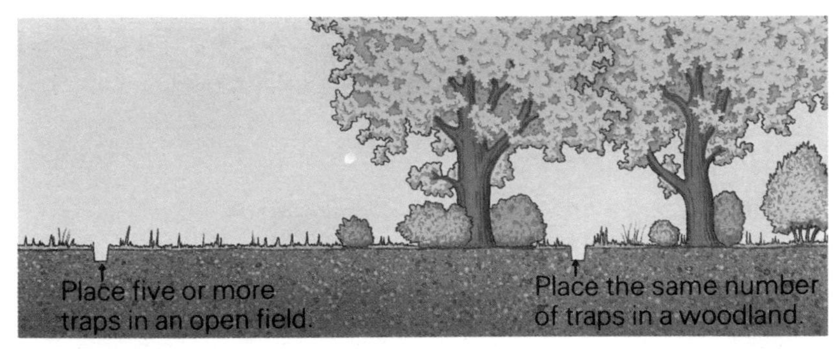

Siting pitfall traps.

2

a Draw two bar graphs to show the numbers of each type of animal caught in each environment.

b Which is the most common animal in the field? Is it also the most common animal in the wood?

c Why do you think there are *both* more animals *and* more types of animal in the wood than in the field?

Type of animal	Number caught in the field	Number caught in the wood
woodlice	2	11
spiders	2	8
centipedes	0	2
beetles	8	10
millipedes	0	2
Total	12	33

Finding out how many

The number of animals caught in a trap only gives you a very *rough idea* of how many of each type of animal there are in an environment. All the animals of **one type** that live together in the **same environment** make up a **population**. The number of animals in a population can be measured *accurately* by marking captured animals and releasing them. These pictures show you how a survey was carried out to find out the size of a population of ground beetles.

You can't be sure to catch all the animals in a population. You have to allow for the ones you don't catch. A formula called the **Lincoln Index** can be used to work out the total number of beetles in the area.

$$\text{Total number of animals} = \frac{A \times B}{C}$$

A = the number of animals that were marked and released.
B = the total number of animals captured in the second sample.
C = the number of marked animals in the second sample.

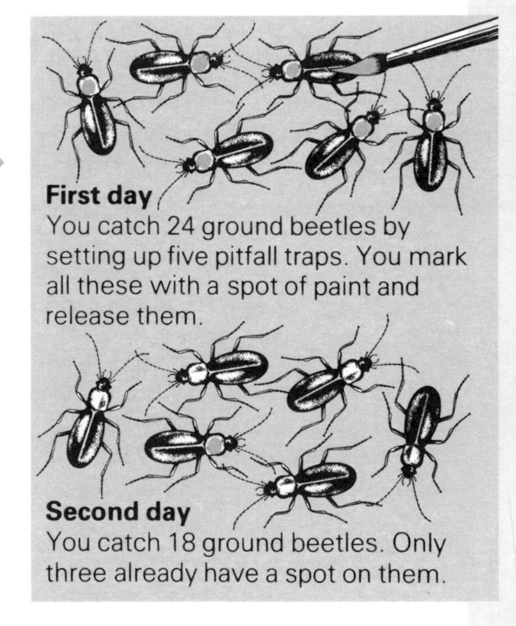

3

a Make a copy of the following table and then fill in the spaces using the information from the diagram above.

b Use the Lincoln Index to calculate the total size of the ground beetle population.

Number of ground beetles collected and marked in the first sample. (A)	
Number of beetles collected in the second sample. (B)	
Number of marked beetles in the third sample. (C)	

4 What is a population? What is a community? Are they the same?

1.11 Hedgerows – an ideal home?

Where to live?

All animals and plants need a suitable habitat to survive. They need somewhere to live just like you do. Where will you find most animals and plants in Britain? This pie chart provides the answer.

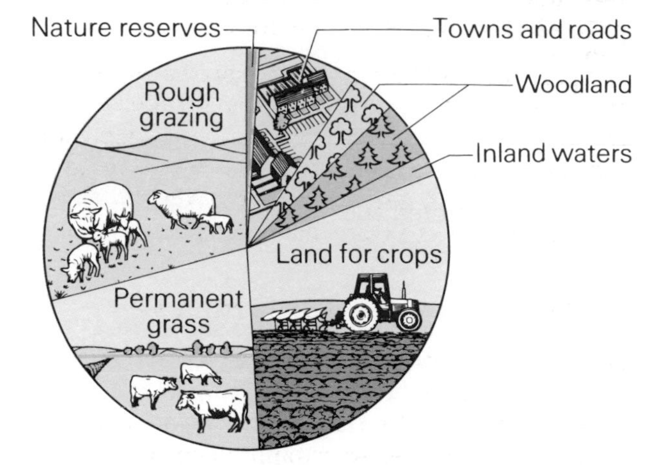

How is land used in Britain?

Living with farmers

The pie chart shows that more land in Britain is used for farming than for any other purpose. About 80% of the land is used for agriculture. Nature reserves which were set up to protect wildlife only take up less than 1% of the land surface. Most wildlife depends on farmers for protection because wildlife often lives on land used for farming.

A farm with hedgerows and traditional-sized fields - suitable for cows or sheep.

Hedged in

The hedgerow is a farmland environment that takes up a large amount of land space. One survey shows that there are over 600 000 miles (1 million km) of hedgerow in the whole of Britain. For hundreds of years, hedgerows like those shown here have been used as land boundaries. They show whose land stops where. They also stop farm animals wandering off, and provide them with shelter from the weather.

Who needs hedgerows?

Not just the farmers! In Britain hedgerows provide the ideal breeding sites for a quarter of all types of insects, two-thirds of all types of birds and three-quarters of all types of mice, hedgehogs and similar animals.

Somewhere to hide and something to eat

Hedgerows provide a habitat that is very similar to the edges of woodlands. This is a very rich habitat containing many types of plants and animals. Many of the animals found in this country depend on hedges as **breeding sites**. You can see why hedges are so important by looking at this picture.

If the hedgerow habitat disappeared, many animals and plants would disappear with it.

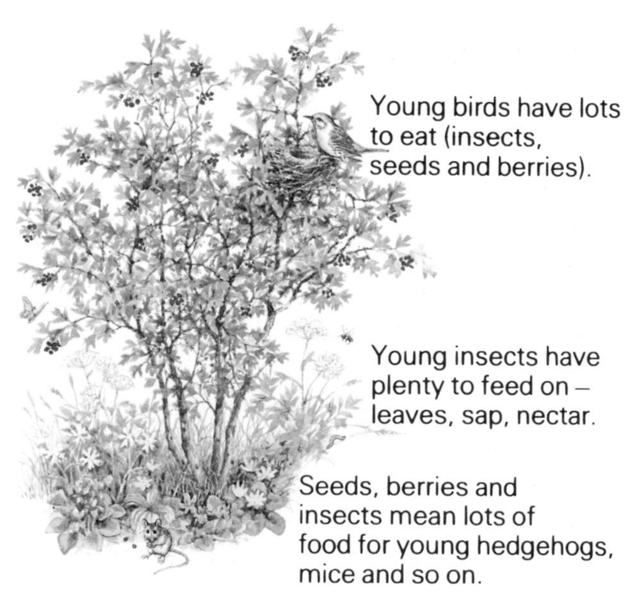

Young animals need a place to hide, a place for their nest and lots of food. Hedgerows have it all!

The unpopular hedge

The fields in the photograph on this page are in the same area as the fields in the picture on page 22. Can you see any difference?

In the 1950's and 60's many hedgerows were removed – in fact an average of 5000 miles (8000 km) of hedgerows were removed each year during this time. This changed the countryside as you can see in the photograph. Why did farmers remove so many hedgerows?

A modern farm with large fields – suitable only for growing crops.

Why remove hedgerows?

Hedgerows get in the way of large farm machinery like the combine-harvester. Some hedgerows were also dug up and removed when houses and factories were built on farmland. Other reasons for removing hedgerows can be seen in the picture of a cornfield below.

Why doesn't the corn close to the hedge grow as well as the rest of the crop?

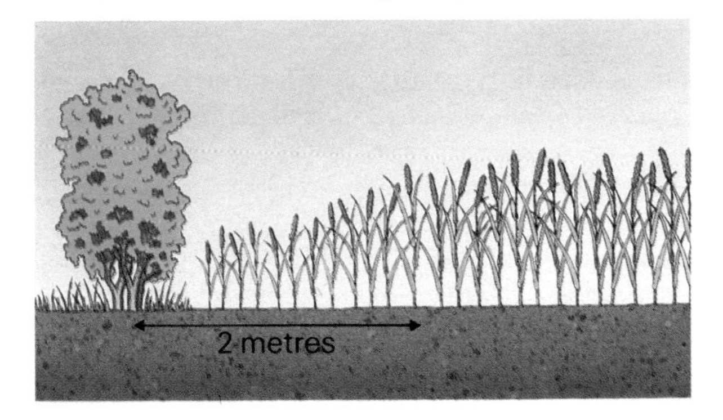

The soil near the hedge has been treated in the same way as the rest of the field. It has had just as much fertiliser and seed added to it. But the corn next to the hedgerow doesn't grow well.

The hedgerow plants take nutrients from the soil that could be used by the growing corn. The hedge also blocks out some of the sunlight. This means the corn close to it cannot make as much food by photosynthesis. Farm pests like fieldmice hide and nest in the hedge. They then feed on the nearby corn as it grows.

The future of farmland wildlife

Modern farms try to use as much land as possible to grow crops. This has a very bad effect on wildlife. About 80% of the different types of birds and 95% of the different types of butterfly will disappear from an area if their habitat is destroyed. Now that farms produce vast amounts of food, much more effort is being made to **protect** wildlife by **preserving** farmland habitats. Hedgerows are no longer being removed – in fact they are being re-planted on many farms.

1. In Britain, which type of farming uses the most land – growing crops or raising cows and sheep?
2. Why were hedgerows planted originally. Give as many reasons as you can.
3. Name three animals that make use of hedgerows. For each example explain how the animal uses the hedgerow habitat.
4. The main use of farmland in East Anglia is to grow wheat. In Somerset the main use of farmland is to keep cows. More hedgerows have been removed in East Anglia. Why?

1.12 The disappearing otter

The private life of the otter

"Tarka pursued one until he caught it, but as he was swimming to the bank he saw another, and followed it with the fish in his mouth. He snicked it as it darted back past his shoulder. Strokes of the heavy tapering rudder over two inches wide at its base and thirteen inches long, that could stun a fish by its blow, enabled him to turn his body in water almost as quickly as on land." (From *Tarka the Otter*)

Otters live in nests called **holts** along the banks of rivers. The holts are usually gaps between tree roots. Reeds and bushes give the otter lots of places to hide. The fish in the river, particularly the eels, are their main source of food.

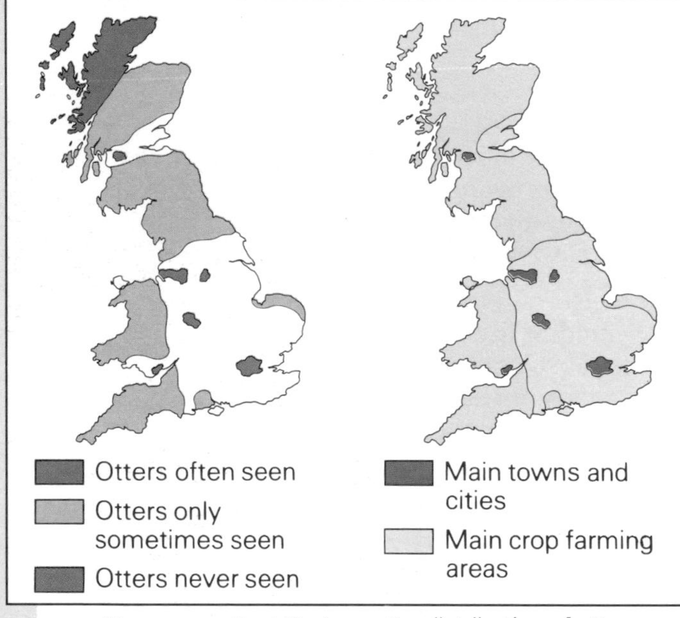

The map on the left shows the distribution of otters. The other map shows the main crop-farming areas.

Where can otters be seen?

From 1977 to 1979 a survey was carried out to find out how many otters there were in different parts of the country. The results are shown on the first map.

Otters used to be a fairly common sight in many areas of Britain. Their numbers declined in the late 1950's and early 1960's. The decline was unusual because it happened at about the same time across certain parts of the country. Some people claimed that the main crop farming areas were the worst affected. Look at both the maps. Do you agree?

The first map shows that otters are now totally absent or very rare in most regions. Only in the remote parts of Scotland has the otter population remained at a high level. This suggests that the presence of people may have caused the decline in the otter population.

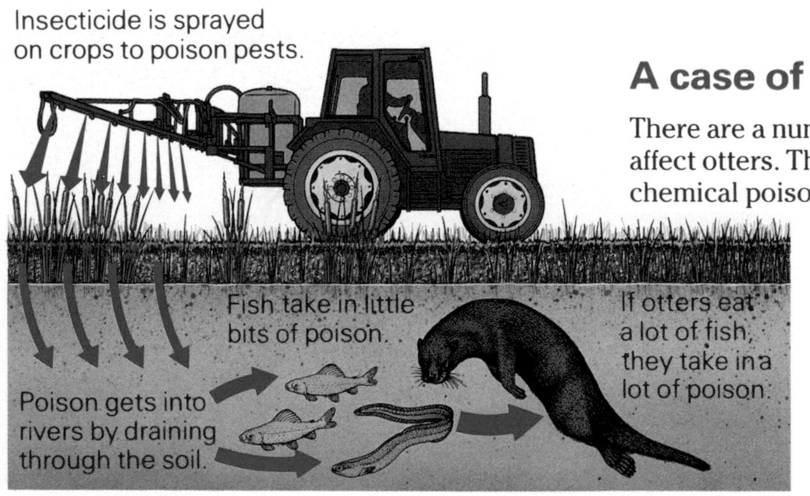

A case of food poisoning

There are a number of ways that people could affect otters. The most likely is the use of chemical poisons called **insecticides** that are used to kill insects. It was in the early 1950's that insecticides such as 'DDT' were first widely used by farmers to kill insect pests. These chemicals were soon washed into rivers and lakes. This diagram shows you the effect these poisons have on the life of the river.

Food chains – a fatal link?

The effect of chemical poisons is often only seen in the final link in a food chain. The concentration of insecticide in the water may not be high enough to kill fish. But over the years, it does poison carnivores such as the otter. Birds like the grebe are also affected.

Just how does the poison become more deadly as it passes along the food chain? This diagram shows how even a little insecticide can harm the otter.

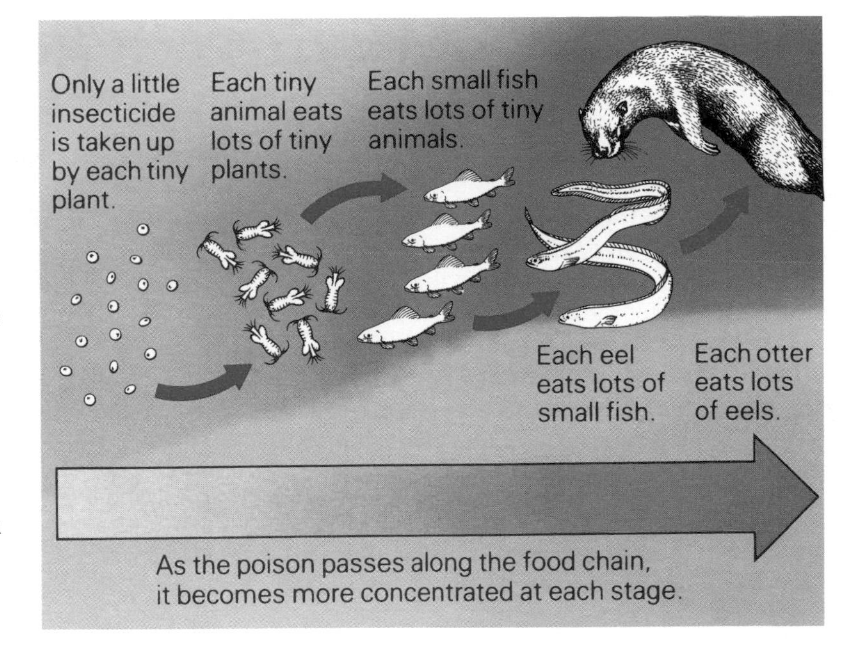

Destroying the otter's habitat

Another reason why the numbers of otters have fallen is the destruction of their habitat. The ideal habitat for otters is where there are reeds, bushes and trees growing on the riverbank. But if overgrown, these plants can block the river and cause flooding. To prevent this the Water Authorities clear away the bushes and trees from the banks so that the river can flow freely. This means that otters can no longer find somewhere to make holts. Also there will be no places where they can hide away during daylight.

This mechanical digger is clearing the overgrown banks of a stream. How will this affect the otter?

Problems with the mink

The mink is a North American animal that was bred on farms here for its fur – mink coats used to be very fashionable. Some minks escaped in the 1930's and now breed in the wild. They feed on fish and make their dens in riverbanks. What effect could this have on otters?

The mink and the otter are different animals but they need similar places to build their dens. They also eat the same food. They are in **competition** with each other for food and somewhere to live. This is another reason why otter numbers remain low. It is an example of how people can sometimes upset life in natural environments by introducing animals from other places.

A mink coming out from its nest under a tree root. The otter would like to live here too!

1 What kind of habitat do otters need to survive?

2 What are insecticides? Why do farmers use insecticides?

3 What is meant by 'competition'? How do minks and otters compete with each other?

4
- **a** When did farmers first begin to use insecticides like DDT?
- **b** When and where did otter numbers decline?
- **c** Explain the gap between the two dates.

1.13 Swan death – who's to blame?

Counting the Queen's swans

For 400 years a regular count has been made of the number of swans on the river Thames near Reading. These birds belong to the Queen. Once a year, at a ceremony called "swan upping", the number of swans are counted. Their bills are notched to show they belong to the Crown. A separate count is made of the cygnets (young swans). The number of cygnets gives a measure of how fast the swans are breeding.

The bar chart shows the numbers of swans counted in 1956 and in 1984. The number of swans and the number of cygnets are shown for each year.

Rounding up the swans on the Thames – the first stage of 'swan upping'.

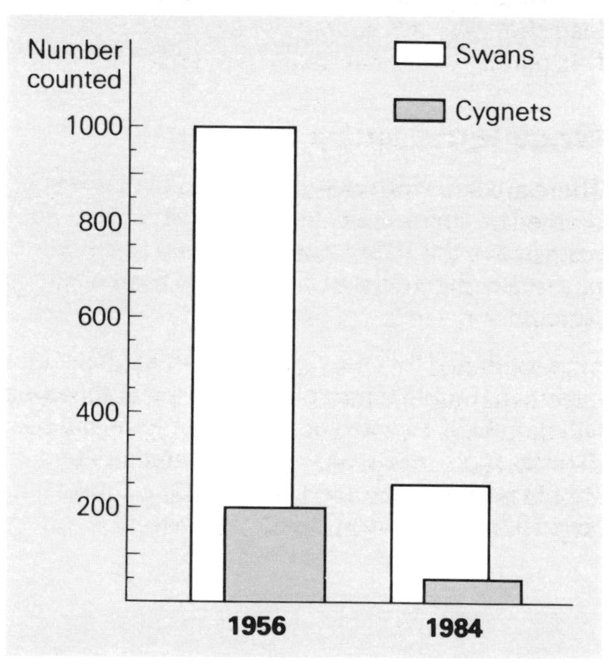

The results of 'swan upping' in 1956 and 1984.

1 How many swans were counted
a in 1956?
b in 1984?

2 **a** There were lots of swans on the river in 1956. What must have happened to them?
b There were less swans present in 1984. Why does this mean that there will be less cygnets in 1984 than in 1956?

3 What does the number of cygnets tell you about the possible numbers of swans in the future?

Why the swans were dying

Swans were dying in many other rivers as well as the Thames. To find out why so many were dying, dead swans from different regions were collected and examined carefully. This table shows the results of the study.

Cause of death	**Region**				
	Thames	Trent	Avon	Northern England	Wales
Lead poisoning	43	12	7	12	8
Collisions with overhead wires	5	3	1	0	0
Internal injuries	5	0	2	1	0
Infection	2	2	0	2	2
Shot or killed	1	0	0	0	0
Cause unknown	9	1	0	0	1

4 **a** What was the cause of most of the deaths?
b How many swans died in this way?

Careless fishing kills swans

Scientists at the Veterinary Investigation Centre in Leicestershire carried out further investigations to find out why swans were dying. When the stomachs of 365 dead swans were examined, 211 contained anglers' lead shot. Other items of fishing tackle such as floats, hooks, and spinners were also found. One swan had a 10 cm long float in its throat! It had made a hole through the bird's throat and penetrated its heart. Lost fishing tackle is a serious problem for swans – but lead shot is the most common cause of death. Why? Because of the way swans feed!

Swans are herbivores. They feed on many types of water weed found on the bottom of rivers and lakes. This is why they are so often seen stretching their necks down to the bottom of the river. They also take up gravel from the bottom to help them to digest their food. The small stones help to grind up the water weed in the swan's stomach.

Anglers' lead shot can be swallowed with the gravel. The lead shot dissolves in the stomach and lead passes into the swan's bloodstream. If lead gets into an animal's blood it will become very ill or even die from lead poisoning.

The swan's special method of feeding means that lead shot affects them but does not affect the fish, plants or other animals in the river.

Some anglers use a float and lead shot to keep hooks at the right level in the water. Very useful to anglers, but deadly to swans.

How to save swans

There are a number of ways to stop swans being harmed by careless fishing. A new law has banned the sale of lead shot for use in fishing tackle. But much lead shot was sold before the ban and may still be in use.

Anglers should be made more aware of the effects they are having on swans. With luck, they will then be more careful when buying and using their tackle. Some people even say that fishing should be banned altogether on waters where large numbers of swans are being killed.

5 Write a newspaper article about the problems anglers are causing for swans. Suggest how these problems can be overcome, but try not to upset the anglers – after all, the swans need the anglers' help!

Bottom feeding!

1.14 Destroying natural habitats

A change of scenery

The pictures below both show sand dune environments. Look carefully at the two pictures. One of the environments has been affected by holiday-makers walking over dunes. What has happened as a result?

This sand dune environment is in a nature reserve.

This sand dune environment is between a holiday camp and a beach.

Trampled to death

The sand dunes in the nature reserve are covered by a plant called marram grass. The dunes near the holiday camp look different – they have large areas of exposed sand. Also there will often be large gaps between each dune. Why do the two areas show these differences?

When people go to the beach from the camp they often trample on the marram grass growing on the dunes. This kills the marram grass and leaves areas of exposed sand. When the maram grass has gone, the wind blows the sand away causing large holes to appear in the dunes.

As more marram grass gets destroyed by trampling, more of the dunes become eroded (worn away) by the wind. The sand dune environment depends on plants such as marram grass to hold the sand in place.

Planting maram grass on bare sand dunes helps to stop the sand from blowing away. Fences protect the young maram grass from trampling.

Disappearing toads

It is not just the dunes that are affected by holiday-makers. Many dunes have damp areas close to them. This land is often used to build holiday homes or caravan parks. The ground has to be drained to make it ready for building. The draining has a big effect on the local wildlife.

This damp and sandy environment is a special kind of habitat of many animals and plants. For example, the natterjack toad can only breed in warm shallow pools found there.

The natterjack used to be common on many coasts but it is now so rare that is a **protected species**. This means that it is an offence to kill, injure or even handle the protected animal (or plant) without a special licence.

The natterjack has become a rare species because much of its habitat has been destroyed. Forestry, farming and holiday sites are all to blame. You may enjoy your holidays by the sea, but the natterjacks don't!

A natterjack toad – the best sites for its young to develop are in warm shallow pools in sandy soil.

Protecting wildlife

The effect of holiday makers on the sand dunes shows how easy it is to change an environment. Once the environment changes it may no longer suit the animals and plants living there so they 'disappear'. The lucky ones may be able to populate a similar environment nearby. But many will just die – particularly the plants. This is why many people work to protect environments against change.

In Britain, the Nature Conservancy Council works to conserve the environment. Many countries have created **national parks** to protect the most important environments.

There are ten national parks in England and Wales and five in Scotland. The parks are areas of the countryside which have special rules to prevent their environment from being changed. These parks even attract a lot of holiday-makers! If they follow the rules, such people need not be a threat to the environment.

Take your litter home

Guard against all risk of fire

Keep to public paths

Fasten all gates

Help to keep all water clean

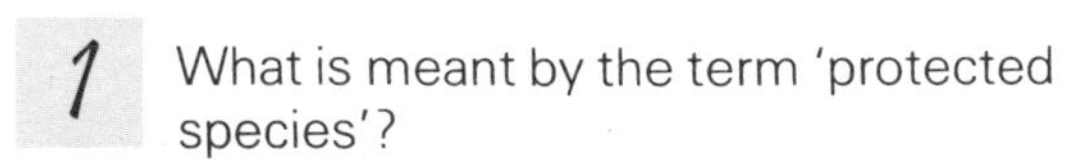

1 What is meant by the term 'protected species'?

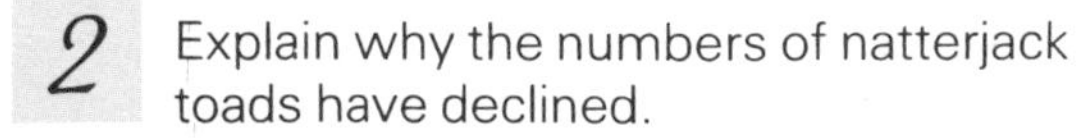

2 Explain why the numbers of natterjack toads have declined.

3 Have you ever seen signs in parks or gardens which say 'keep off the grass'? Why should you do what they tell you?

4 Find out which national park is nearest to you. What special features of the environment are conserved there?

1.15 Decisions affecting the environment

Little Upton is a small quiet town in the country. The main talking point among the town's residents is the news that a new factory is to be built close to the town. The factory will need a site near a **main road** so that lorries can deliver raw materials. Most of the products will be sent out by **rail**. The map below shows where the factory may be built.

People's opinions about the proposed building are reported in the town's local newspaper shown on the opposite page. Read the article carefully to grasp the different views.

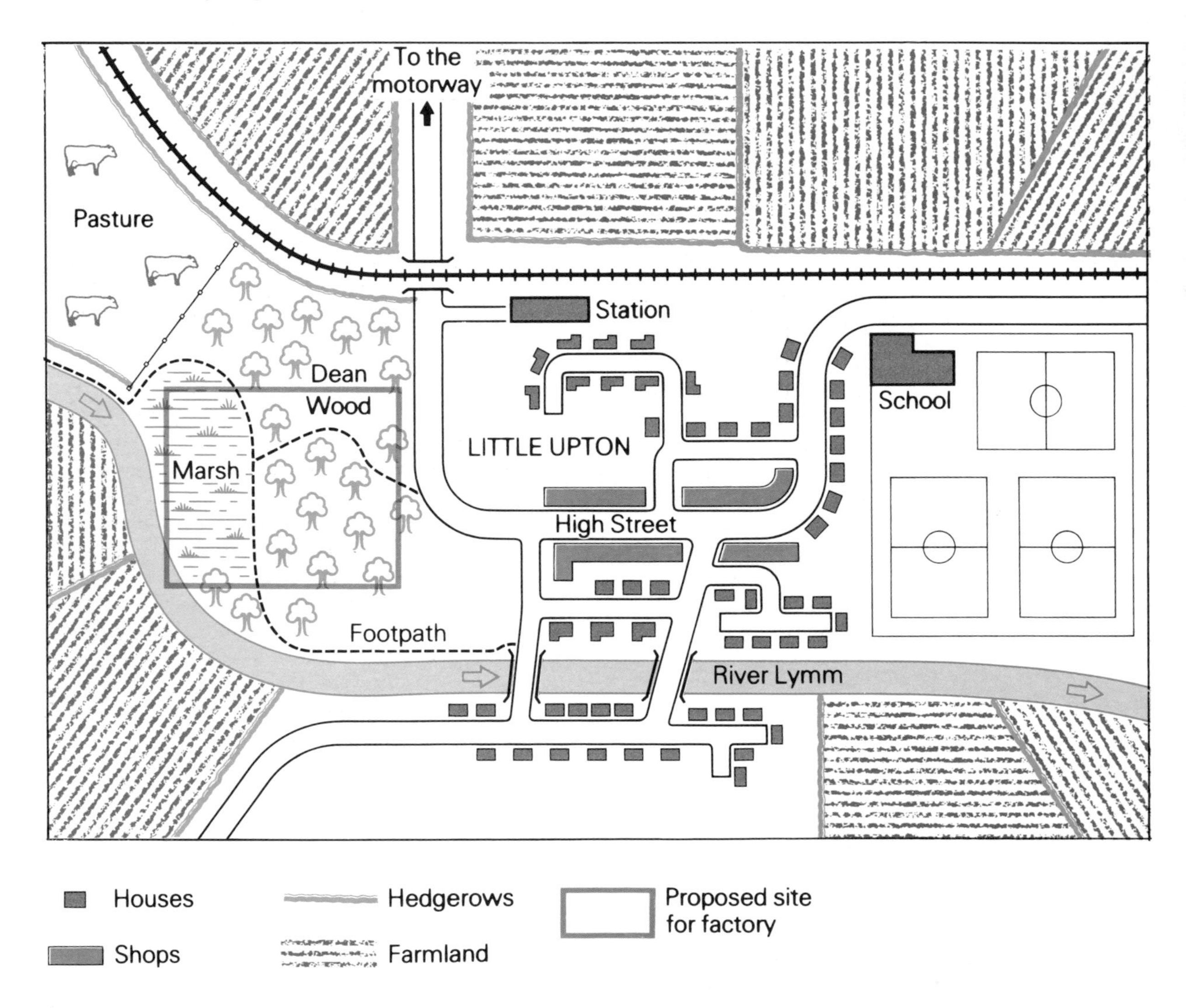

UPTON JOURNAL

Little Upton's local weekly newspaper

Factory site starts row

Talks have been taking place this week between the local council and the owners of Fleet Foods. The large food processing company plans to build a new factory on land at Dean Wood.

MORE JOBS

Negotiations are still going on but council leader Peter Jones said, "We're optimistic about the outcome."

He added, "The factory will bring many jobs to the area. People have been forced to leave the area to look for work elsewhere. This will keep Little Upton alive.

Other members of the council agree, Councillor Delia Walker said, "The proposed land is of no value. People use it to dump rubbish. Even the local farmer can't grow anything there because it keeps flooding." She went on, "The only other site that can be used is the playing fields and this is of great value to the community."

BEST SITE

Councillor Jones added a warning to the people of Little Upton, "If the factory is built next to the school, lorries will be going back and forth along the High Street. My biggest fear is the danger of lorries passing the school all day.'

The owner of Fleet Foods, Mr Sam Wheatley, appealed direct to the public, "I hope everyone realises that the Dean Wood site involves little or no disruption to the town." He continued, "Sadly not everyone agrees with our proposals but the council is doing its utmost to make sure that this great opportunity is not lost."

LOCAL CONCERN

Some residents are angry with the council's proposals. Edith Molyneux, aged 60, said, "It will be a shame to spoil such a beautiful area. I walk along the river and through the wood as often as I can. Why do they have to spoil this part the town?"

Another concerned resident, Mrs Ann Watson, lives along the river. She asked, "What if the factory starts to pollute the river? All its muck will flow through the town. Who will be responsible for clearing up any mess?"

THREAT TO WILDLIFE

Jim Pickering a member of the RSPB states, "The marsh and the wood have a very rich birdlife, Redshanks breed on the marshland. These rare birds will disappear if the area is disturbed."

He added, "The wood contains tawny owls and green woodpeckers. Why can't the owners of the factory find an alternative site?"

Other people agree. Among them is the Little Upton Conservation Group. Their spokesman Mr. Fred Doyle said, "We have been trying to clear up rubbish dumped in the wood and we have fenced off part of the marsh to keep the farmers' cattle out. These areas are really important wildlife habitats. Why don't the council build on the playing fields? Does the school really need three playing fields for the kids?"

1

a For the Dean Wood site, make a list of:
  i its advantages;
  ii its disadvantages.

b Do the same for the playing field site.

c Imagine you have the final decision. Which of the two sites would you choose? Why?

2 Can you see any other site which may be suitable for the factory? Is this other site already in use? Why would it be suitable?

3 Fleet Foods are to buy part of the pasture and let it grow wild. This is to help replace Dean Wood. How long will it take for the new wood to grow? Is there any other habitat that could support the wildlife while the new wood grows?

4 The farmers of Little Upton are thinking of removing the hedgerows from their fields to make large fields. How will this affect the wild plants and animals in the area?

2.1 What is pollution?

The environment in which you live contains many different organisms. One way or another, these organisms depend on each other and the world around them in order to stay alive. All living organisms produce waste materials – often this waste is used by plants so they too can stay alive. Only a few waste materials can be used in this way. Most other forms of waste damage the environment – this is what is meant by **pollution**.

Most pollution is caused by humans, but there are a few other sources of pollution.

How can pollution be harmful?

If you throw away a chocolate wrapper, you may be harming the environment. In Great Britain, nearly 1 million tonnes of rubbish are produced each week. A lot of that rubbish does not end up in the dustbin. It is left lying around, polluting the environment.

Other types of pollution can be deadly. If waste from certain factories gets into the rivers, many animals will be poisoned. Poisonous gases from cars and chimnies can spread through the air and kill trees in forests thousands of miles away.

Even when only a few people drop litter, it soon builds up.

The brown spots were caused by the chemical pollution which killed these fish.

'Dead wood' – poisoned by long-range air pollution.

What a waste!

Power stations produce the electricity that we all like to use – but they also produce waste. Coal is burnt in power stations to make electricity. As it burns, the coal releases poisonous gases into the air. This waste can pollute the environment.

The process of making electricity involves heating large quantities of water. This hot water is often pumped into rivers while it is still warm. This waste heat then kills many organisms that live in the rivers. Some countries do not let this hot water go to waste. Can you think how the hot water could be used?

There are many other types of pollution which could be avoided if the waste was put to good use.

Pollution can be hard to detect – there's no smoke, but the river may still suffer.

It's the tall thin chimneys of power stations that pollute the air.

If this field gets polluted, the poison can pass from the plants to the cows and to you!

What needs to be done?

Be more aware of your environment – watch out for things that might pollute it. Remember that pollution is harmful to the many organisms living there – including you!

What things should you look out for? Thick dust and smoke are a sure sign of air pollution. Oily water or dead fish in a stream show water pollution. But only a few types of pollution are easy to see. Much dangerous waste goes unnoticed – until it is too late.

This module will help you to find out more about pollution – its causes, its effects and what to do to stop it.

2.2 Take care of the air

What is air pollution?

If smoke pollutes the air, it is easy to see. This is because it contains black particles of carbon which are known as soot. Other pollutants are invisible gases like sulphur dioxide and carbon monoxide. You can't see them, so the air might look clean, but they can poison you.

Not all invisible gases are poisonous, but they can still be pollutants – carbon dioxide is one example. Air pollution can be due to many different pollutants. In each case it means that the air contains some chemical which is not normally found there.

Air pollution – making a mess of the air you breathe.

Where does it come from?

Fuels like coal, oil and petrol all produce air pollution when burnt.

How much pollution is there?

Often there only needs to be a tiny amount of a pollutant to make the air polluted. The level of pollution is found by measuring the mass of pollutant present in a standard volume of air. The mass of the pollutant is measured in **microgrammes** (μg) – one millionth of a gramme. The standard volume of air is **one cubic metre.**

These measurements can be used to compare one area with another. The amount of air pollution can vary quite a lot from area to area.

Air Pollutant	Amount of air pollutant in microgrammes per cubic metre of air	
	Area A	Area B
sulphur dioxide	80	25
lead	1300	50
smoke	110	25
nitrogen dioxide	321	119
carbon monoxide	106	33

Which area would you prefer to live in?

Air pollution and health

You are continually breathing air into your lungs. This means your lungs are the most likely part of your body to suffer from the effects of air pollution. Bronchitis is a type of lung disease – the numbers of deaths it causes are shown in the top graph.

Can you see a connection between the deaths from bronchitis and the distance from a city centre?

The bottom graph shows that the closer you get to the city, the greater the amount of sulphur dioxide pollution in the air. Some people have suggested that there is a connection between the deaths caused by bronchitis and the level of sulphur dioxide pollution. Do you agree with this idea?

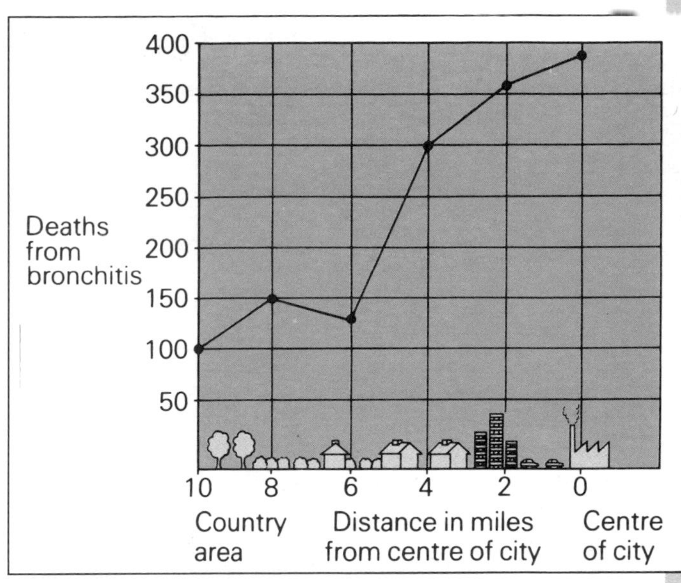

Where is the greatest risk of suffering bronchitis?

Other pollutants

There are many poisonous pollutants in the air – such as mercury and cadmium from industrial sources, lead from car fumes. These can get from your lungs into your blood. These pollutants build up slowly until they reach poisonous levels. They are called **cumulative poisons.**

Prevention is the only cure!

Nowadays, the amount of smoke produced when burning fuels has been greatly reduced. For example, natural gas is a smokeless fuel used in many homes and factories. Solid fuels are also available which produce very little smoke. But they are more expensive than coal which produces sooty smoke. This is because they have to be treated in a special way. There is also lead-free petrol on sale. Although it is cheap to produce, it cannot be used in all types of cars.

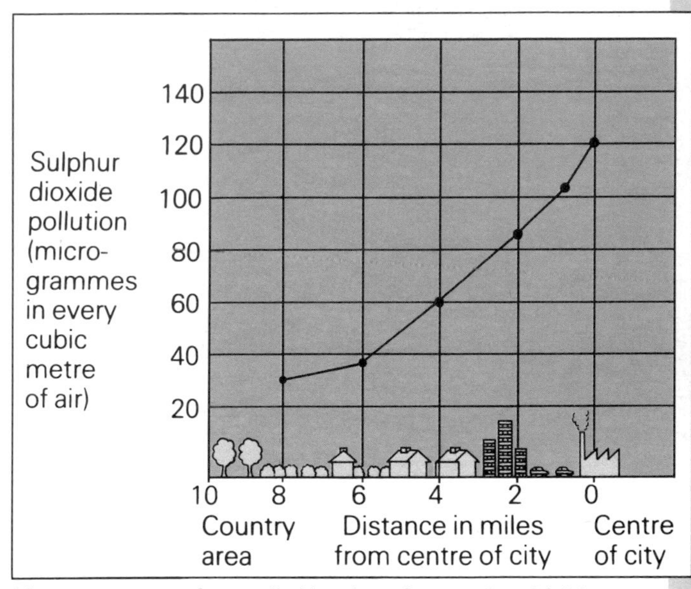

Hot waste gases from tall chimnies often cool and fall in quite short distances.

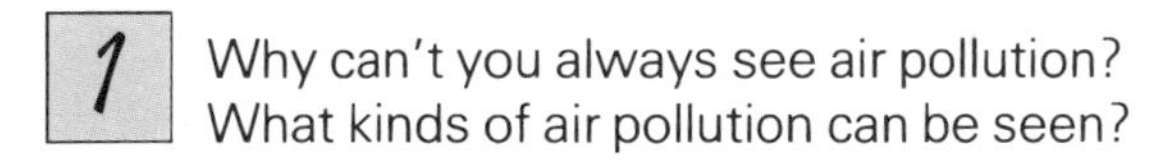

1 Why can't you always see air pollution? What kinds of air pollution can be seen?

2 Name two poisonous gases that are air pollutants.

3 Look at the table opposite:
- **a** What units are used to measure air pollution?
- **b** Which pollutant is found in the largest amounts? What causes it?
- **c** Which area (A or B) is nearer to a city? Give a reason for your answer.

4
- **a** Name two fuels which cause sulphur dioxide pollution.
- **b** Give two examples of cumulative poisons that are air pollutants.
- **c** How do these poisons get into your body?

5
- **a** What is unusual about the number of bronchitis deaths 4 miles from the centre of the city?
- **b** Can you suggest a reason for this? (*Hint*: what goes up, must come down!)

2.3 Acid rain, a deadly downpour

'Vinegar rain'

The vinegar you put on your fish and chips tastes bitter. It is a **weak acid** – a chemical which turns indicator paper red. But you don't just find weak acids in vinegar bottles – you can find them in the air too!

Once there was a rainstorm in the USA with rain that was 1000 times more acidic than vinegar. This type of rain is known as **acid rain**.

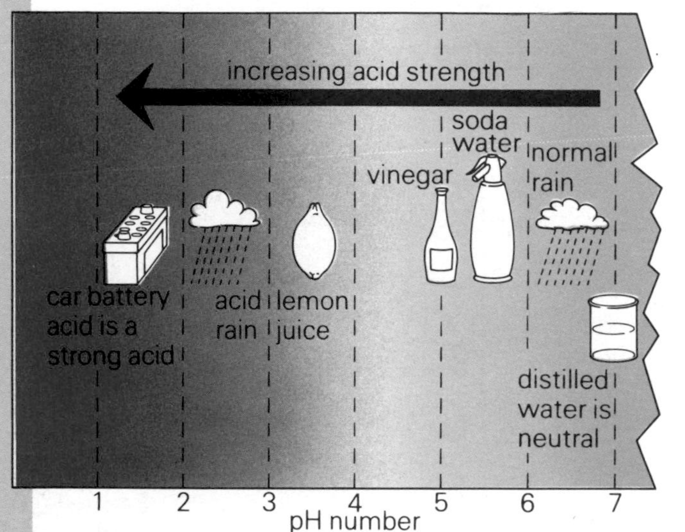

The lower the pH value, the stronger the acid.

What goes up, must come down

Every time you switch on the TV, you are helping to produce acid rain. The electricity you use is produced in power stations by burning fuels such as coal and oil. Travelling in cars or buses causes acid rain too. This is because of the fuel burnt in the engine.

When these fuels are burnt, gases such as sulphur dioxide, nitrogen dioxide and carbon dioxide are produced. If these gases dissolve in rainwater in the air, acid rain is formed. The strengths of different acids can be compared using a special scale called the **pH scale**. Vinegar has pH number of about 5. Distilled water is not at all acidic – it is *neutral*, with a pH of 7.

Eating away

What do the pictures below tell you about the effect acid rain has on stone? Some metals such as iron and zinc are used on the outside of many buildings. These metals are also affected by acid rain – they lose their shine and become weak and brittle.

On the left is a stone figure on the wall of Lincoln Cathedral. On the right is the same figure 70 years later.

Innocent victims

If acid rain eats away at stone and rock, what must it be doing to living things? Large amounts of acid can kill. When acid rain falls, it collects in rivers and lakes, making the water acidic. All the living things in some lakes have been killed by acid rain. The small animals and plants in the water die first. Then the fish die too – and so do all the other animals, which feed on them!

Trees also suffer from the effects of acid rain. In Europe, large areas of forest are dead or dying as a result of acid rain.

Should power stations avoid burning fuels which produce acid rain?

Curing the problem

Efforts have already been made to make lakes less acidic. One way of doing this is to add lime to the lake water. Lime is an **alkali** – a chemical which reacts with acids and makes them weaker. When lime is added, the lake water becomes less acidic and its pH increases. If more acid rain falls into the lake, more lime must be added.

The best way to prevent acid rain is to stop the gases which cause it from escaping into the air. In power stations, this is now done by passing all the waste gases through an alkaline solution. This is called **scrubbing** and removes the gases which cause acid rain. Car exhausts can be fitted with **converters** which change nitrogen dioxide (one cause of acid rain) into harmless nitrogen.

These methods of limiting air pollution are quite effective, but may make electricity and cars more expensive. A choice may have to be made – either to save money or else to help to save the environment.

This truck is spraying lime powder into a lake to reduce the acidity caused by acid rain.

1. What is the pH number for lemon juice?
2. Which is the stronger acid – lemon juice or acid rain?
3. Otters feed mainly on fish. Why would otters be affected by acid rain?
4. If you want to stop acid rain, why is "scrubbing" gases from a power station better than adding lime to lake water?
5. How would animal life in a forest be affected by acid rain?

2.4 Nature's warning signs

Are you breathing clean air?

The air you are breathing may contain particles or gases that can damage your health. But how can you tell? In Tokyo, the pollution is so bad that there are large electronic scoreboards in public areas. These scoreboards show people the amount of pollution in the air they are breathing!

Watch out for pollution!

There are simpler, natural ways of detecting pollution. Plants called **lichens** can show how much pollution there is in the air.

Bushy lichens need really clean air. They are easily poisoned by just a small amount of sulphur dioxide. If you have this lichen growing near you the air will probably be very clean.

Leafy lichens can put up with a small amount of air pollution. If you can find this lichen growing on tree bark, the air is probably quite clean.

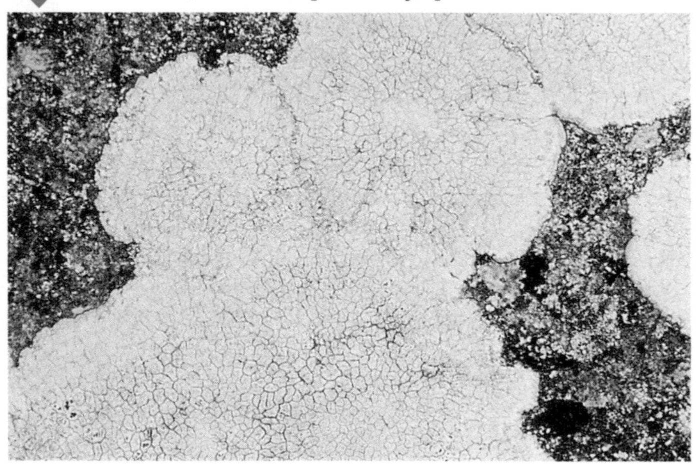

Crusty lichens are able to survive in air polluted with sulphur dioxide. They can be found growing on trees and walls in most town centres.

If there are *no lichens* around the air must be heavily polluted with sulphur dioxide.

A tough life

Lichens grow in very exposed places. Not many plants can grow on the dry surface of rocks or tree bark. To grow in such places, lichens need to be very good at **absorbing** water and nutrients. They get these by absorbing any rainwater which falls on them. Rainwater contains very small amount of nutrients which are just enough to keep the lichens alive.

Taking in poison

When the air is polluted, rainwater contains more than just small amounts of nutrients. It also contains poisonous pollutants, such as sulphur dioxide. For many lichens, absorbing polluted rainwater is the same as absorbing a poison.

Some lichens can stand quite a high level of sulphur dioxide, but others are poisoned when there is just a trace of the gas in the air. This means that the presence of lichens can provide you with a clue of how much pollution there is in the air – they are **pollution indicators**.

Studying pollution

This map shows the results of a study of lichens found in a particular area. Study the map carefully and then answer the questions below.

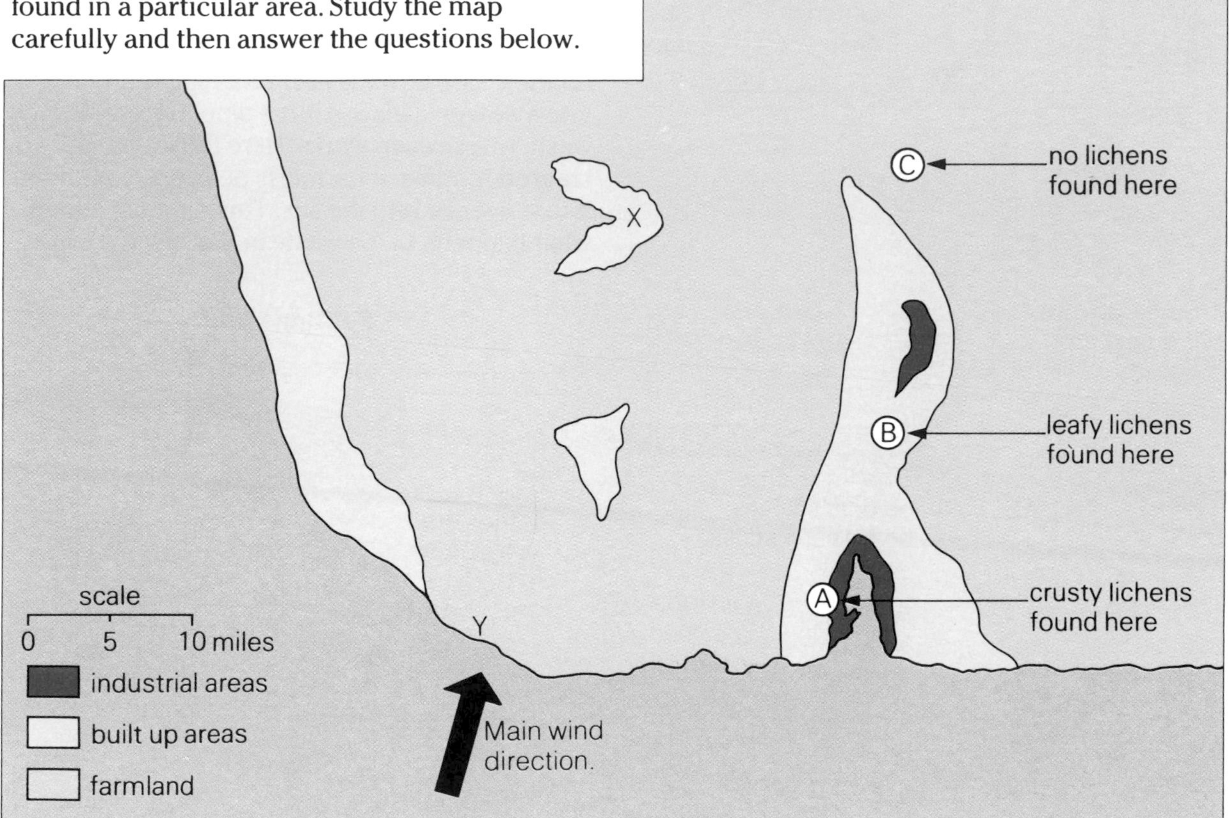

1 What sort of lichen grows only in areas of almost no pollution?

2 How do lichens get enough water and nutrients to grow in very dry places?

3 **a** Which area on the map above (A, B or C) shows the highest level of air pollution? Give a reason for your answer.
b Even though the two industrial areas produce the same amount of pollution, the pollution in this area is higher than elsewhere. Why do you think this is so?

4 What lichens would you expect to find in
a region X
b region Y?
In each case explain your answer.

5 When the sulphur dioxide pollution in an area was reduced, it took years before the lichens growing in the area showed any change. What reasons could you suggest for this delay?

2.5 Mucking about with water

Pulling the plug out

What would your life be like if you had no water supply to your home? You would soon realise just how essential water is in your daily life. You use it for drinking and cooking. You wash yourself and your clothes in it and you use it to flush the loo. The water that has been used becomes smelly, filthy and full of harmful bacteria. When you pull the plug out or flush the loo, where does all this filthy waste go?

Fighting the filth

All the waste water is flushed from your home into a **sewer**. This is a large pipe that carries the waste to a sewage works. Here the waste is **treated** to make it harmless before it is pumped into a river or into the sea. This diagram shows what happens to the waste in the sewage works.

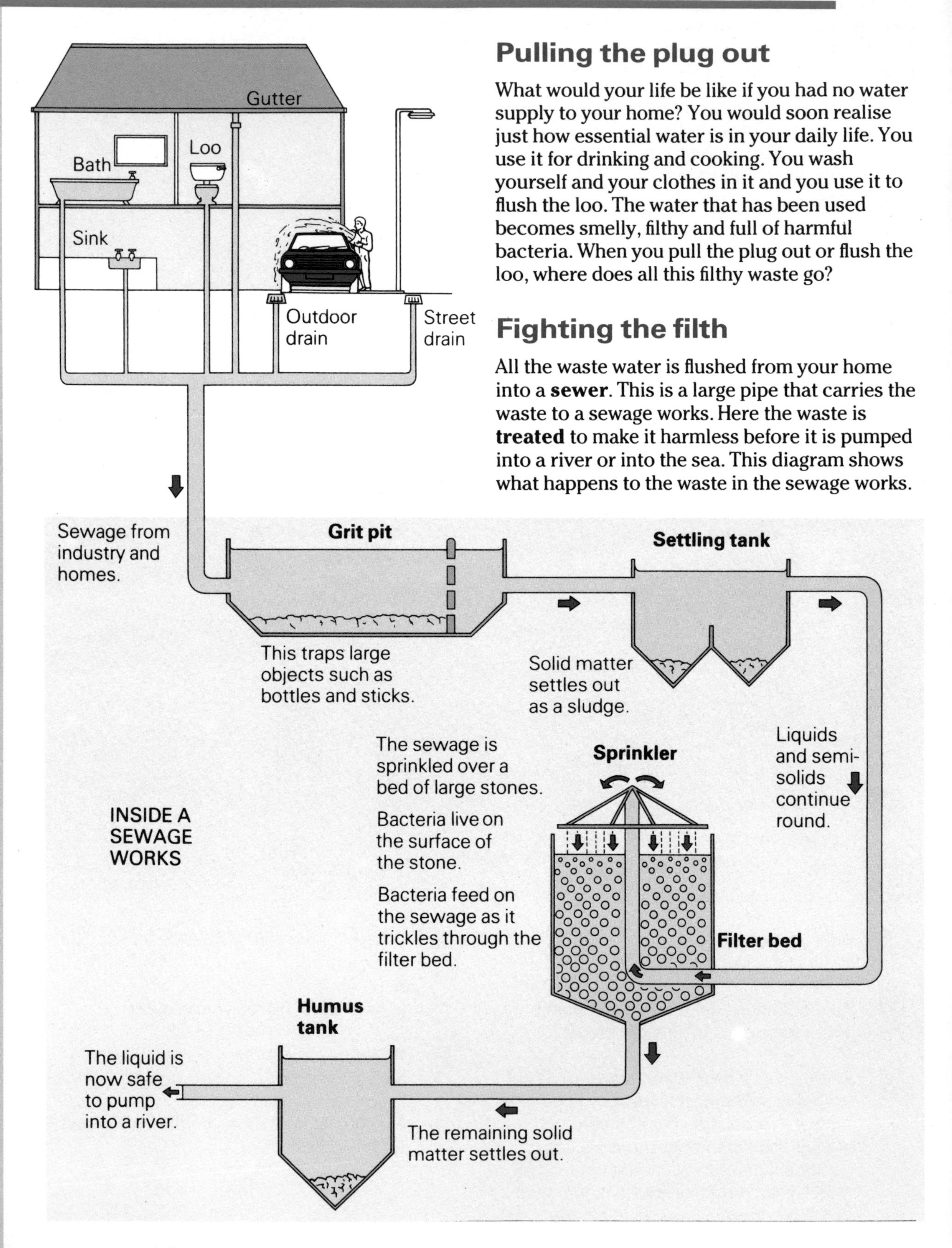

Preventing disease

Untreated sewage may contain bacteria that can cause serious diseases such as cholera or typhoid. In this country, sewage is made harmless before it is poured into a river. The treatment of sewage prevents the spread of such diseases by killing the bacteria that cause the disease. In many countries, human waste (faeces and urine) is put straight into rivers. The same river may be used to provide drinking water for many people. When the waste of one person suffering from typhoid gets into river water, hundreds of other people can become infected.

Diseases can spread rapidly if drinking water is polluted with untreated sewage.

Food for bacteria – death to fish

As well as being a health hazard, untreated sewage can also damage the environment. When untreated sewage is put into a river, it becomes food for **decomposing bacteria**. These are bacteria that obtain their energy from dead or waste material. To release energy from sewage, the bacteria need oxygen. The effect this has on the river can be seen in this diagram. ▼

Clean water containing plenty of oxygen and a wide variety of animals.

Bacteria use up oxygen as they break down the sewage. Fish and many other animals die because of a shortage of oxygen.

Sewage continues to be broken down as the river carries it along. It may be carried a long way downstream before the river returns to normal.

1. Name four main stages in the treatment of sewage.
2. Name two diseases that can be spread by untreated sewage.
3. Sometimes sewage contains poisonous pollutant substances produced by industry. What effect could this have on the working of the filter bed?
4. In some sewage works, air is blown through the sewage after it has passed through the settling tanks. Why does this help to treat the sewage?
5. Why do fish die when untreated sewage is added to rivers?

2.6 Murky waters

This river is obviously polluted but it is not always so easy to tell.

Testing for pollution

How can you tell that the water in a river is polluted? The water may look clean but it could contain harmful substances that will kill all the life in the water. The most common substance that pollutes rivers is **sewage**.

Sewage can kill aquatic life by removing oxygen from the water (for more details, see pages 40 and 41). The oxygen is used up by bacteria which feed on the sewage. If a river contains only a small amount of oxygen, it is a sign that it may be polluted by sewage. The amount of oxygen can be measured using an oxygen meter.

Pollution clues

You can also use the presence of certain animals to tell you about the state of a river. When the amount of oxygen falls, some animals will swim away and others will even die. Only a few animals that can live in low levels of oxygen will stay. The animals that can be found indicate whether the river is clean or polluted. They are **indicator animals**.

A similar lack of oxygen is caused by the waste from paper factories and food factories. The bacteria can also feed on this waste and again cause the oxygen level to fall.

A selection of indicator animals.

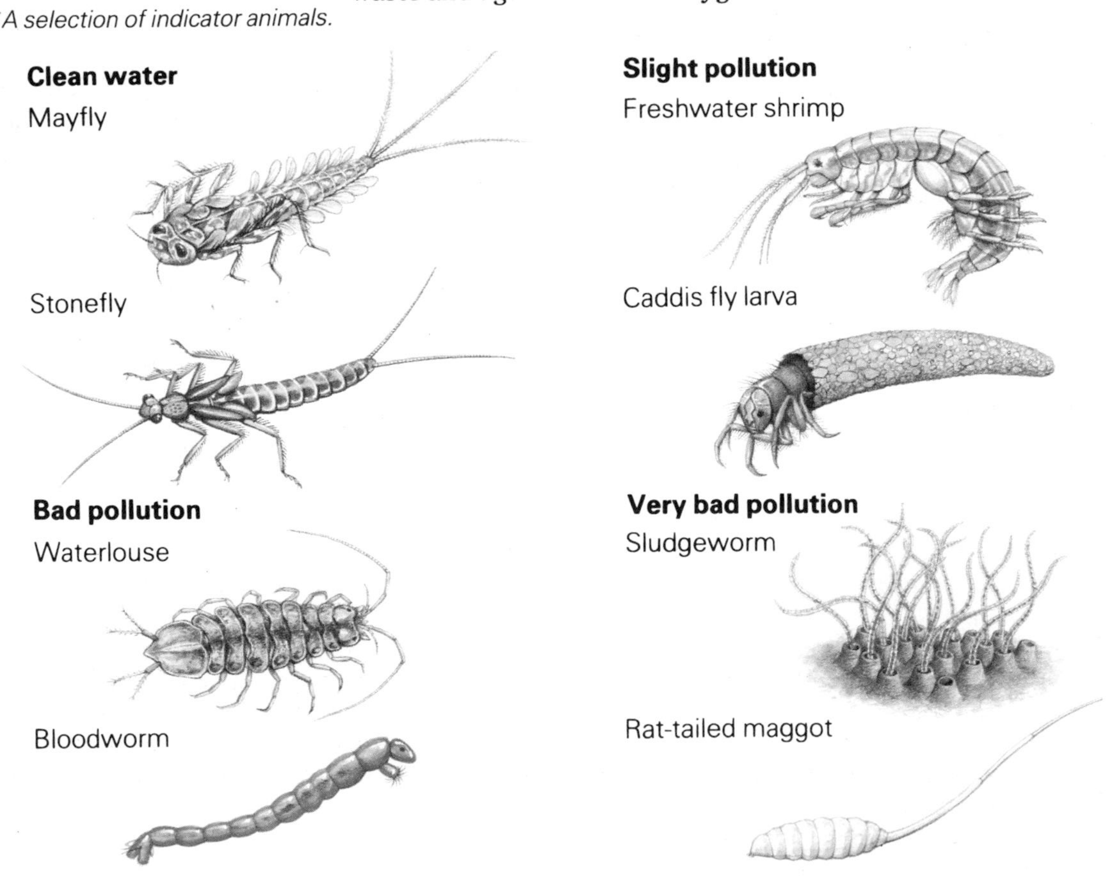

What's more ...

There are many other pollutants which simply poison the water directly. Some of these come from modern farming methods. Others are due to waste from homes and factories.

Phosphates and nitrates are chemicals which are used as fertilisers to help crops grow. Before long, rainwater washes them into streams where they poison the animal and plant life. While solving one problem in farming, the fertilisers help to create a pollution problem!

Many soaps also contain types of phosphates. As well as 'bars' of soap, these phosphates are found in washing-up liquid, washing powder and shampoos. The sewage system is not always able to remove these soaps and they end up polluting the rivers and streams.

Engine oil, diesel and petrol are very damaging pollutants. These substances should *never* be poured down a drain. Even spilling them on the soil can cause problems to the water supply. Most councils have a special arrangement for people who want to get rid of old oil and fuel.

Fertilisers increase plant growth – and water pollution!

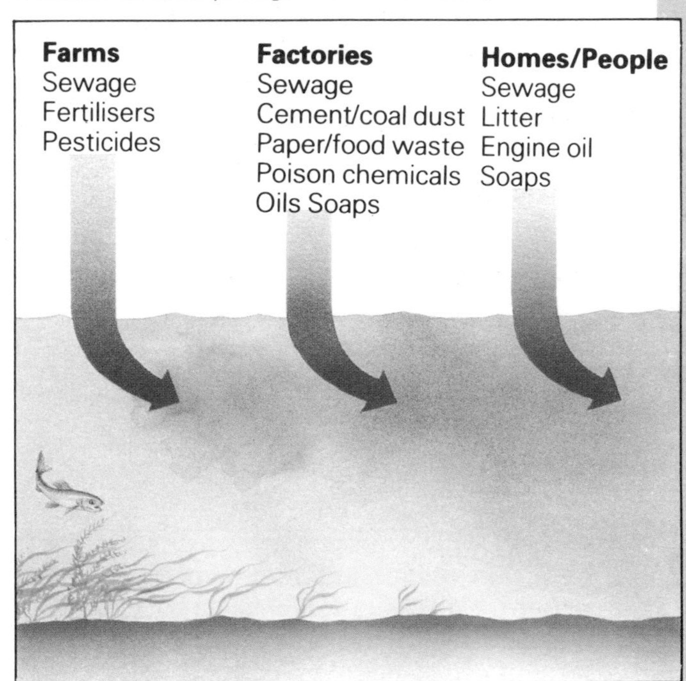

Fighting water pollution involves controlling all these threats – and many more.

Keeping it all in check

There are strict controls which stop factories from polluting the sewage system. Scientists from Water Authorities have to keep a careful check on river pollution. There's a lot to watch out for! Poisonous metals such as arsenic and nickel; dust such as cement powder – in addition to all the sewage, oil, litter, nitrates and phosphates.

Anyone caught polluting the rivers or sewage system will have to pay a large fine. Over many years, the threat of being fined by the Water Authority has helped to stop people from polluting the water supply.

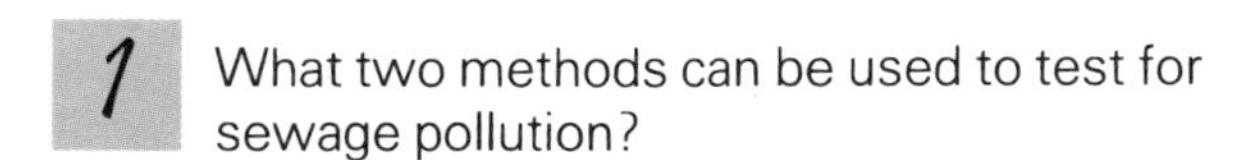

1 What two methods can be used to test for sewage pollution?

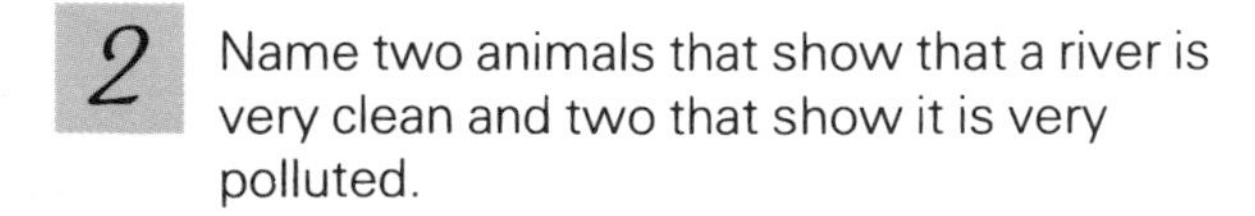

2 Name two animals that show that a river is very clean and two that show it is very polluted.

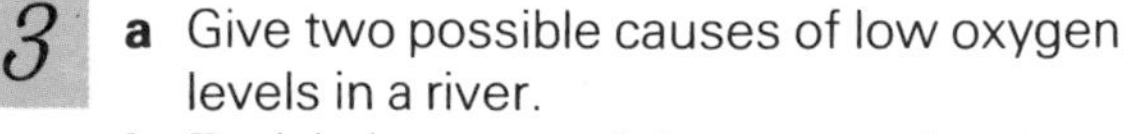

3
- **a** Give two possible causes of low oxygen levels in a river.
- **b** Explain how *one* of these actually reduces the oxygen level.

4 Describe two ways in which phosphates pollute rivers and streams.

5 Suppose a river contains a lot of mayflies and stoneflies at point A. Downstream at point B, there were only bloodworms and sludgeworms. What could have caused this difference? Explain your answer as fully as possible.

2.7 Protecting life in the sea

Oil means troubled waters

Oil is a major water pollutant. It is lighter than water and so floats on the surface. Oil pollution at sea is very harmful to sea-birds like the Guillemot in the photograph. When sea-birds dive into polluted sea-water to catch fish they become coated in a layer of oil. The oil clogs their feathers so that they cannot fly. The damaged feathers can no longer keep the birds warm and many die from the cold. Others are poisoned by the oil because they swallow it as they try to clean themselves.

Where does the oil come from?

Huge quantities of oil are transported all around the world in very large tankers. If one of these tankers has an accident, large slicks of oil soon spread across the surface of the sea. Oil slicks can have a very damaging effect on the sea and sea shore.

Four ways to clean up

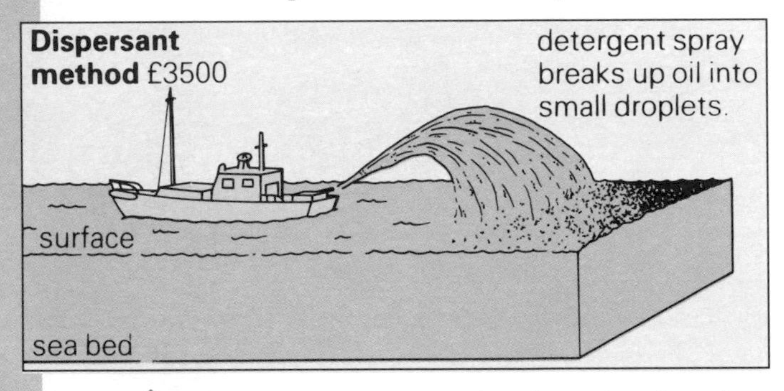

▲ The **dispersant method** needs a few days before the oil clears. It is efficient but expensive. The detergents may poison fish and other animals and plants.

The **boom and skimmer method** gathers the oil up using a long barrier called a boom. A skimmer then scoops the oil off the surface into a suitable container. ▼

▲ The **sinking method** is quick and protects birds and beaches. However the oil pollutes the sea-bed and may even get washed ashore.

Shore cleaning is a last-stop measure. The sea has already been polluted and detergents and mechanical shovels are used to clean the beach.

Large scale pollution

An oil slick can be very difficult to control. It moves with the wind and the tides and can threaten a whole coastline.

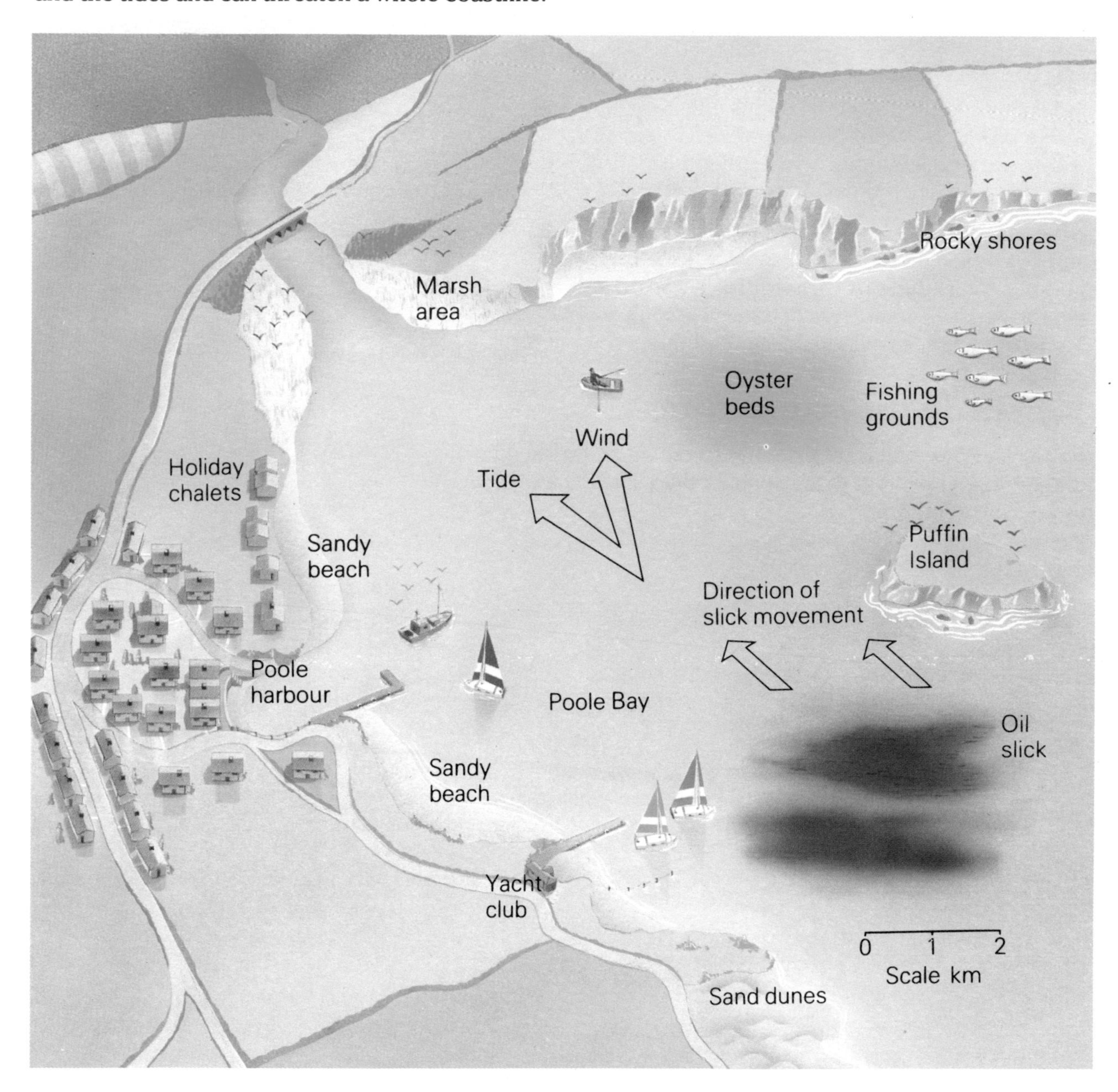

1 **a** Why is floating oil so dangerous to seabirds?
b How does oil kill them?

2 Why does oil float on the surface of the water?

3 **a** Make a list of the areas around Poole Bay that will be affected by the oil slick.
b Explain how oil will affect each area.

4 **a** Which method of control would you use to protect the birds nesting on Puffin Island?
b What problems could this method of control have in other areas?

5 The pollution control officer has £4500 to spend to protect the whole of Poole Bay. Explain how you would spend this money to protect the Bay in the best way possible – don't ignore the disadvantages.

2.8 Manufacturing pollution

Tall chimneys are used to release waste gases as far off the ground as possible.

What are industrial pollutants?

Industry makes many of the things you need, but it also produces waste. Some factories release **waste particles** into the air as either dust particles or particles of heavy metals.

Other industries release **waste gases** such as sulphur dioxide into the air. Industry also pollutes rivers and tips with **liquid** and **solid wastes** such as detergents and heavy metals. These are all types of industrial pollutants.

Where do they come from?

Power stations, chemical industries, cement factories, quarries and the farming industry are some of the main sources of industrial pollution.

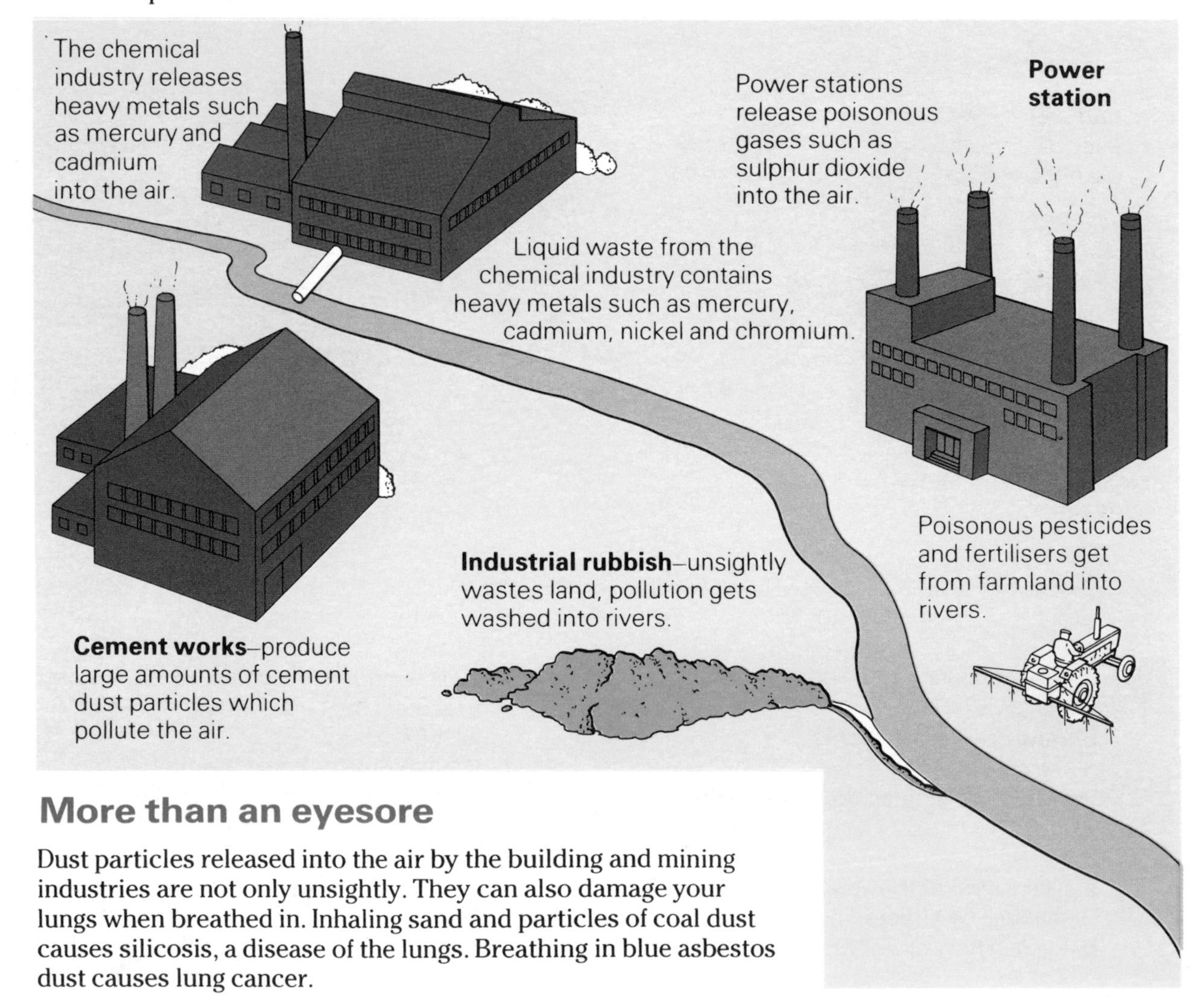

More than an eyesore

Dust particles released into the air by the building and mining industries are not only unsightly. They can also damage your lungs when breathed in. Inhaling sand and particles of coal dust causes silicosis, a disease of the lungs. Breathing in blue asbestos dust causes lung cancer.

Useful but harmful

Heavy metals have many uses in industry. Nickel, chromium and cadmium are used to stop steel from rusting. The largest use of cadmium is in making long lasting nickel-cadmium batteries for the electronics industry.

These metals are **carcinogens** – this means they are chemicals which cause cancer. Mercury is another heavy metal and is used in many industrial processes. If it is breathed in, it builds up in the body and causes liver, kidney and brain damage.

A serious case of mercury poisoning occured in Japan about 20 years ago. Mercury pollution from a factory became concentrated in the flesh of fish in the area. Fish was the main food of the villagers living nearby. The more fish they ate, the more mercury the villagers took into their bodies.

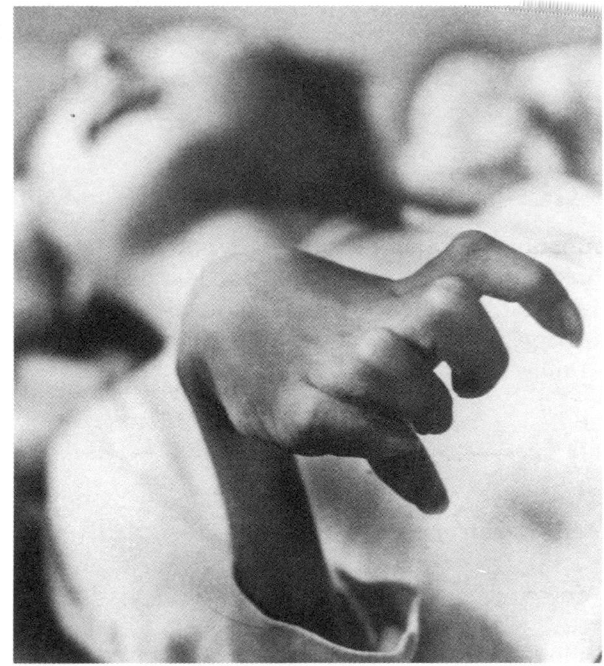

This Japanese child's hand shows a typical deformity caused by mercury poisoning.

Looking out for clean air

One method of monitoring dust pollution is to suck air through filter papers. The light reflected off the filter paper is then measured.

A filter paper with no dust on it reflects 100% of the light. The table shows the light reflected from samples of filter paper at different places.

Heavy metals are not so easy to detect. People working with heavy metals have to be especially careful. Their work area needs to be well ventilated if they are to remain healthy. Any waste gases need to be pumped away from the work place. They can then be treated before being released into the air.

Place	Filter paper	% light reflected from filter paper
(clean filter paper)	○	100
site A	○	96
site B	●	40
site C	○	71
site D	○	76

Measuring dust pollution.

1 Name two sources of industrial pollution.

2 Name two diseases caused by dust particles.

3 Why has cadmium become an increasingly important pollutant in recent years?

4 Give two ways in which mercury can get into the body.

5 **a** Put the results in the table in the form of a bar graph
b Which site is probably located near to a cement works? Give a reason for your answer.

6 Suggest *another* method by which you could measure the amount of dust on the filter paper.

7 What other precautions do you think workers in the heavy metal industry should take?

2.9 Radioactive waste

What is a radioactive substance?

All substances contain energy. A radioactive substance releases some of its energy as invisible **radiation**. This radiation can be made up of *either* fast moving tiny particles *or* rays similar to radio waves.

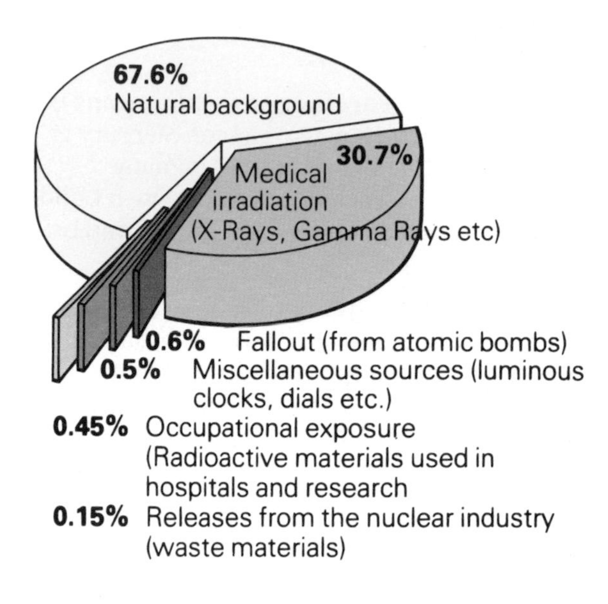

Which is the most common source of radiation you experience?

Is this radiation dangerous?

The energy from radiation can damage living cells. It can cause leukaemia (cancer of the blood) and many other cancers. It can also cause mutations in sex cells which will then lead to babies with deformities.

Your body is exposed to this kind of radiation every day but only to very small amounts. Wood, granite, air and soil all contain some naturally radioactive substances. Sunlight also includes some harmful radiation.

These sources contribute to a **natural background radiation**. The amount of energy from background radiation is very small and the chance of your body being damaged is low.

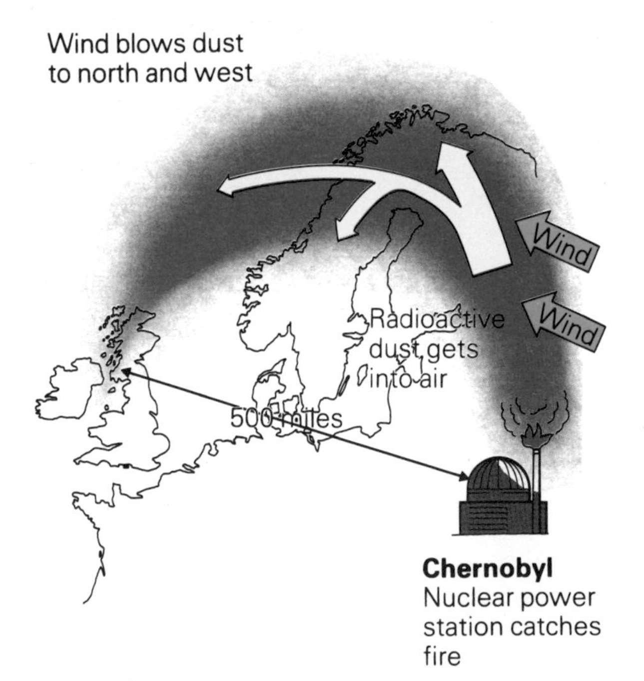

Not much radiation is needed to damage your health – so radioactive pollution from far away can still be a risk.

Increasing the risk

These days, some of the radiation that your body experiences is caused by our technology. Power stations and atomic bombs have produced radioactive waste materials. These have polluted our environment through leaks and fallout from explosions. In some places the level of background radiation may be no longer harmless.

The recent disaster at Chernobyl released massive amounts of radioactive waste into the air. Countries all over the world had to react very quickly to this pollution by checking drinking water and food for radioactive substances.

Even the radioactive sources used in hospitals, industry and research can be dangerous. In Brazil recently, an old hospital x-ray machine was damaged when left at a rubbish tip. Several people were taken seriously ill due to leaks from the machine's radioactive source.

There's always more...

Nuclear power stations continually produce radioactive waste. This has meant large amounts of nuclear waste have built up. These are increasing even faster now because more countries have started to use nuclear power to make the energy they need.

Low level waste include things such as clothing, containers and liquids. Some countries will not allow the large amounts of such low level waste to be dumped in their own country. This waste is often transported and dumped in other countries that do not have strict controls on pollution. Some of this waste is even dumped in the sea. But just getting rid of it does not mean it's safe!

Greenpeace supporters trying to prevent the dumping of nuclear waste at sea.

High energy waste

Some waste is very dangerous and will release large amounts of harmful radiation for thousands of years. This waste must be contained for a great length of time so that it cannot leak out into the environment. The nuclear industry spends vast amounts of money on the transport and storage of this waste in its attempts to prevent high energy waste pollution.

Guns have killed more people than have atomic bombs. Coal power stations continually produce waste gases that pollute the environment. But these facts do not mean that people and the environment are at less risk from nuclear weapons and nuclear power stations. Just because a risk is less common, it does not become less dangerous. Compare driving and flying – car accidents are much more common, but plane crashes are much more deadly.

Radioactive substance	How your body is affected
Strontium	Found in foods such as milk, cream and cheese. Absorbed into the production of bone cells. Linked with the disease leukaemia.
Caesium	Found in plants and animals (lamb, cows) that feed on plants. Linked with certain types of cancer.
Plutonium	An extremely poisonous substance. Tiny particles in the air settle in the lungs. Linked with lung cancer.

How radioactive waste can affect you.

1. You are exposed to radiation that comes from six main sources. What are they?
2. Draw a bar chart to show the amounts of radiation that come from the six main sources of radioactivity.
3. Name two radioactive substances and explain how they can affect your body.
4. The amount of radiation that your body receives each year is steadily increasing. Why do you think this is so?
5. What are the main problems with waste from nuclear power stations?

2.10 Pesticides – A growing threat

The case of the disappearing falcons

Peregrine falcons are magnificent birds of prey which feed on smaller birds, such as pigeons. These falcons attack their prey in a spectacular way, catching it in mid-flight. Peregrines have no natural enemies but their survival was once in doubt. How could this have been?

In the late 1950's, the number of peregrines suddenly decreased. If the number of these birds had kept decreasing, they would soon have disappeared completely. To prevent this from happening, it was necessary to find out why these birds were vanishing.

Peregrine falcons catch their prey in the air, but finish their meal on the ground.

Looking for clues

The scientists who noticed this decline had also made two other important observations. The first was that the number of peregrines had fallen more in some areas than others. In which areas had most of the peregrines died?

Their second observation was that dead pigeons were often found near fields which had been planted with wheat seed. In the 1950's farmers used to soak wheat seed in a chemical poison called **dieldrin**. This was done to prevent the seeds from being eaten by an insect called wheat bulb fly. Why do you think the pigeons were dying?

What effects could these deaths have on the peregrines living in the area? Do any of the scientists' observations support your ideas?

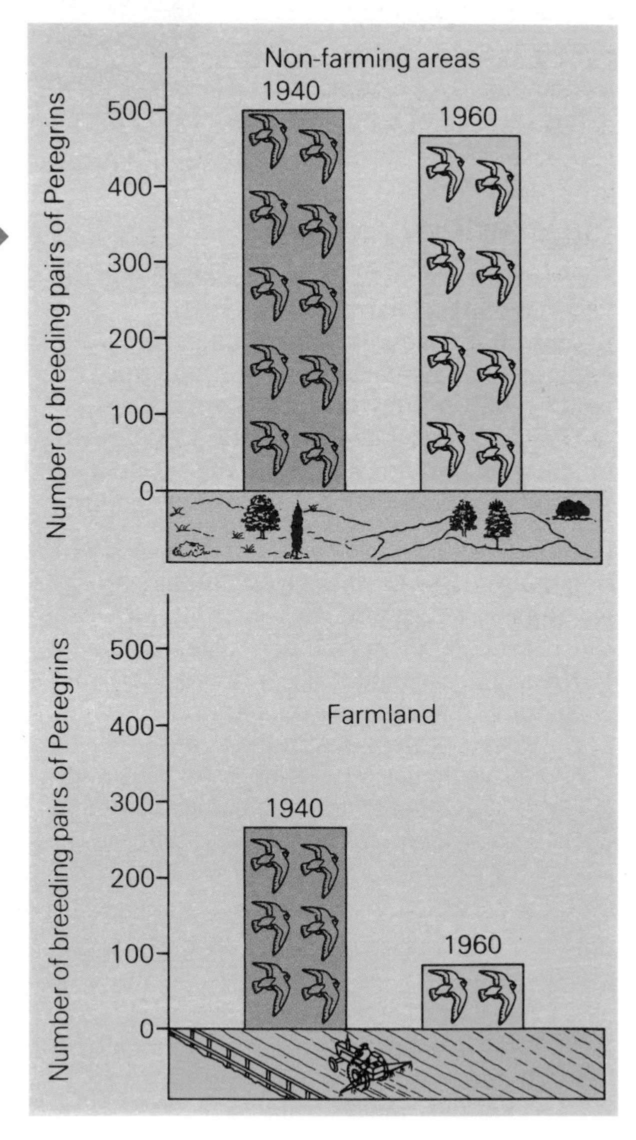

Death by poisoning

Only a little dieldrin was needed on each seed to kill the insect. The amount on one seed was not enough to kill a pigeon. But once a pigeon had eaten a lot of seeds, it would have eaten too much poison and would die. A peregrine suffered in a similar way when it ate poisoned pigeons.

The dieldrin poison even affected the birds long before they died. It made them lay eggs which would not hatch. The poison was killing both the adults and their off-spring at the same time.

Why use chemical poisons?

Every year, large amounts of chemicals are sprayed on the soil or over crops by farmers all over the world. All together, the cost runs into millions of pounds. Many of the chemicals are poisons that kill pests. But why spend so much time and money killing pests?

Pests can do an enormous amount of damage to crops. It has been estimated that a third of the world's food production is destroyed by pests. The crops can be damaged at any of a number of stages – as seeds in the soil, as young plants or even after the crop has been harvested. To protect their crops from pest damage, farmers use chemical poisons called **pesticides**.

Poor growth due to attack by soil pests shows which plants were not treated by pesticide.

War on pests

There are several groups of pests – and often thousands of pests within each group.

Herbicides are chemicals that kill weeds – they are also known as weedkillers. If weeds were allowed to grow they would take away root space, soil nutrients and sunlight from the growing crops.

Fungicides kill tiny organisms called fungi. These organisms grown on the surface of seeds, plants or harvested crops.

Insecticides are used to kill insect pests that eat crops. DDT, dieldrin and pyrethrin are three examples. These are often sprayed over large fields by using special small aeroplanes.

Rodenticides are used to poison rodents such as rats and mice. These pests feed on stored grain and other crops. Warfarin is a common rodenticide – it causes death by internal bleeding.

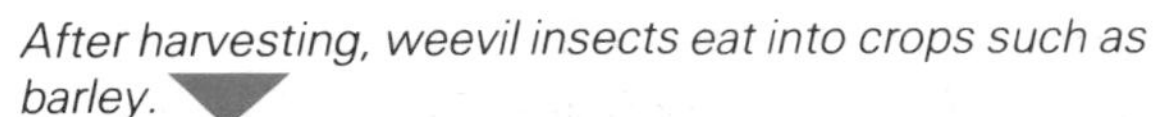

After harvesting, weevil insects eat into crops such as barley.

Careful storage of crops is necessary to prevent pests from ruining food.

1 a What is meant by the terms *rodenticide* and *insecticide*?
b Give one example of each.

2 Name a pesticide that is sometimes an air pollutant. How does it get to pollute the air?

3 a How many breeding pairs of peregrines disappeared from non-farming areas between 1940 and 1960?
b How many disappeared from farming areas over the same period?
c Explain why there was a difference in the two areas.

4 If one pigeon did not contain enough poison to kill a peregrine, explain why the bird of prey was still at risk.

5 One of the pesticides described will be no use in protecting harvested crops. Which one is it? Explain your answer.

2.11 Polluting your body

Smoking isn't glamorous – it's expensive, smelly and dangerous.

Out of sight, but in your body

You can immediately notice some pollutants such as oil slicks, waste heaps and smoke from chimneys. But when you pollute your body you often can't see the harm being done. It can be a long time before you recognise the effect a pollutant is having on your health. By then it is often too late to do anything about it.

It is bad enough for people to suffer from pollution – but some people **choose** to add to this by 'polluting' their bodies through smoking.

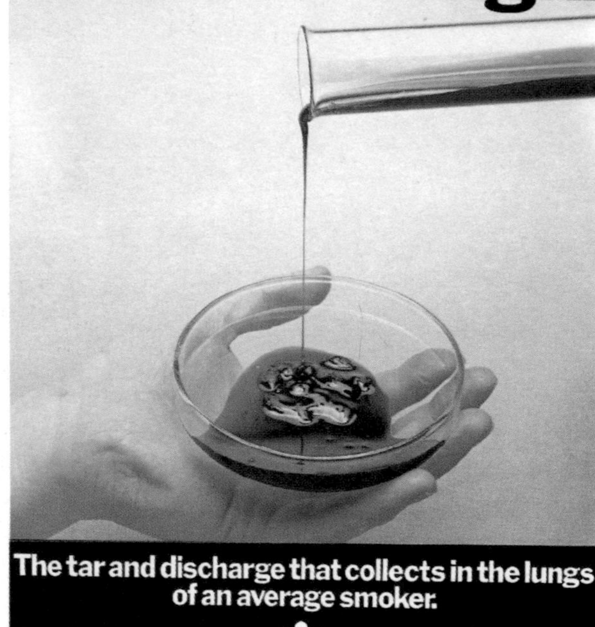

Smoking 20 cigarettes a day means filling your lungs with up to 200cm^3 of tar each year.

What's in a cigarette?

Tobacco in cigarettes contains over 400 different chemicals. The most harmful of these are nicotine and tar. The amount of nicotine and tar in a cigarette depends on the type and brand. Government figures show that one cigarette may contain as much as 25 mg of tar and 2.5 mg of nicotine. When tobacco is burnt, it also produces a very poisonous gas called carbon monoxide – the same gas that is produced by burning petrol in cars!

Smoking can kill you...

Cigarette smoke is made up of sooty particles and tar. As smoke is inhaled, the tar gets trapped in the lungs. Tar damages the lungs by blocking them up. This leads to breathing disorders such as **lung cancer** and bronchitis. The number of deaths from lung cancer increases as more cigarettes are smoked. The chemicals which cause lung cancer are contained in the tar.

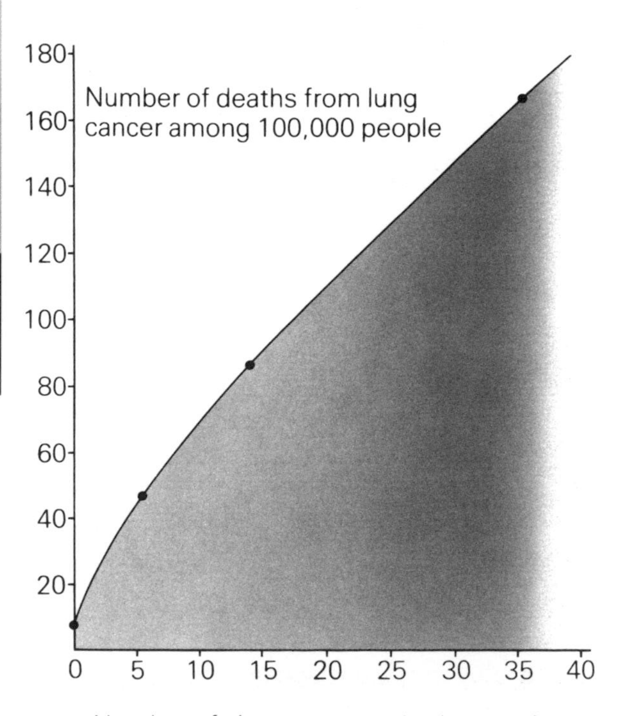

Smoke more, die sooner!

...in more ways than one.

When smoke is inhaled, carbon monoxide and nicotine from the smoke dissolve in the blood in the lungs.

Nicotine is the chemical which makes smokers become addicted to cigarettes. This means that they find it very hard to stop smoking.

Carbon monoxide takes the place of oxygen in the blood and makes smokers short of breath. Carbon monoxide also weakens the heart.

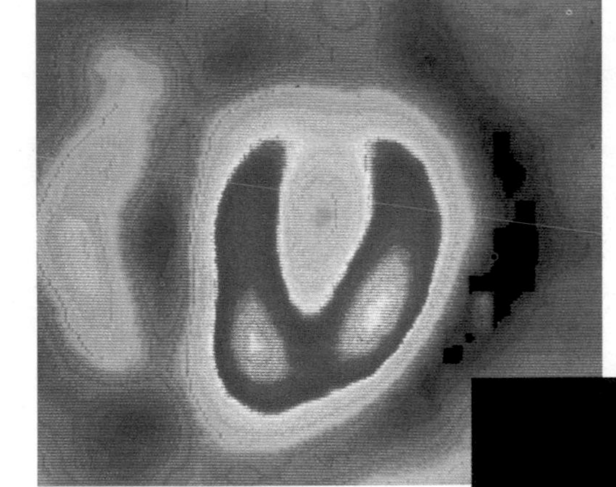

Healthy heart muscle shows up red in this scan. Which is the smoker and which is the non-smoker?

Many smokers die of heart attacks because the carbon monoxide has slowly killed off their heart muscles. It affects the heart most of all for the simple reason that this is the first organ to be supplied with blood from the lungs. Even after giving up smoking, smokers need plenty of exercise to strengthen their hearts again.

By smoking, you can kill others

A smoker 'actively' smokes a cigarette. A nearby non-smoker will also breathe in the smoke but from the air – not direct from the cigarette. The non-smoker is said to be a **passive smoker**.

This may happen regularly to a non-smoker. For example, when smokers and non-smokers share a house or workplace. In these cases, the non-smoker risks suffering from a smoking-related illness – even though he or she may have never smoked a cigarette.

Regular passive smoking makes non-smokers 30% more likely to get lung cancer.

1 How many different chemicals are found in a cigarette?

2 **a** If you smoked 20 cigarettes a day for one year, how much nicotine would you inhale?
b How much money would it have cost you?

3 How many people died from lung cancer who were:
a non-smokers;
b 20 cigarettes a day smokers?

4 **a** What chemical in a cigarette causes lung cancer?
b What chemical makes smokers short of breath and causes heart disease.

5 Why do smokers find it very hard to stop smoking?

6 Do you think smoking should be banned in all public places? Give reasons for your answer.

2.12 Noise pollution

What a racket!

If you've ever been to pop concert or disco you'll know it can be an enjoyable but noisy experience. You may agree that some kinds of music are rubbish – but would you think of it as pollution? While touring in Germany in 1986, a British group was reported to have caused a small earthquake. How about that for affecting the environment?.

Sound is a form of energy but when it becomes so loud that it can harm you, it is called **noise**. Noise can be thought of as a form of pollution because it can damage your body.

'Heavy metal' can mean noise pollution as well as chemical pollution!

Where does it come from?

Life in towns is generally very noisy. How often do you get to hear complete silence? Not very often, there is always some noise to be heard.

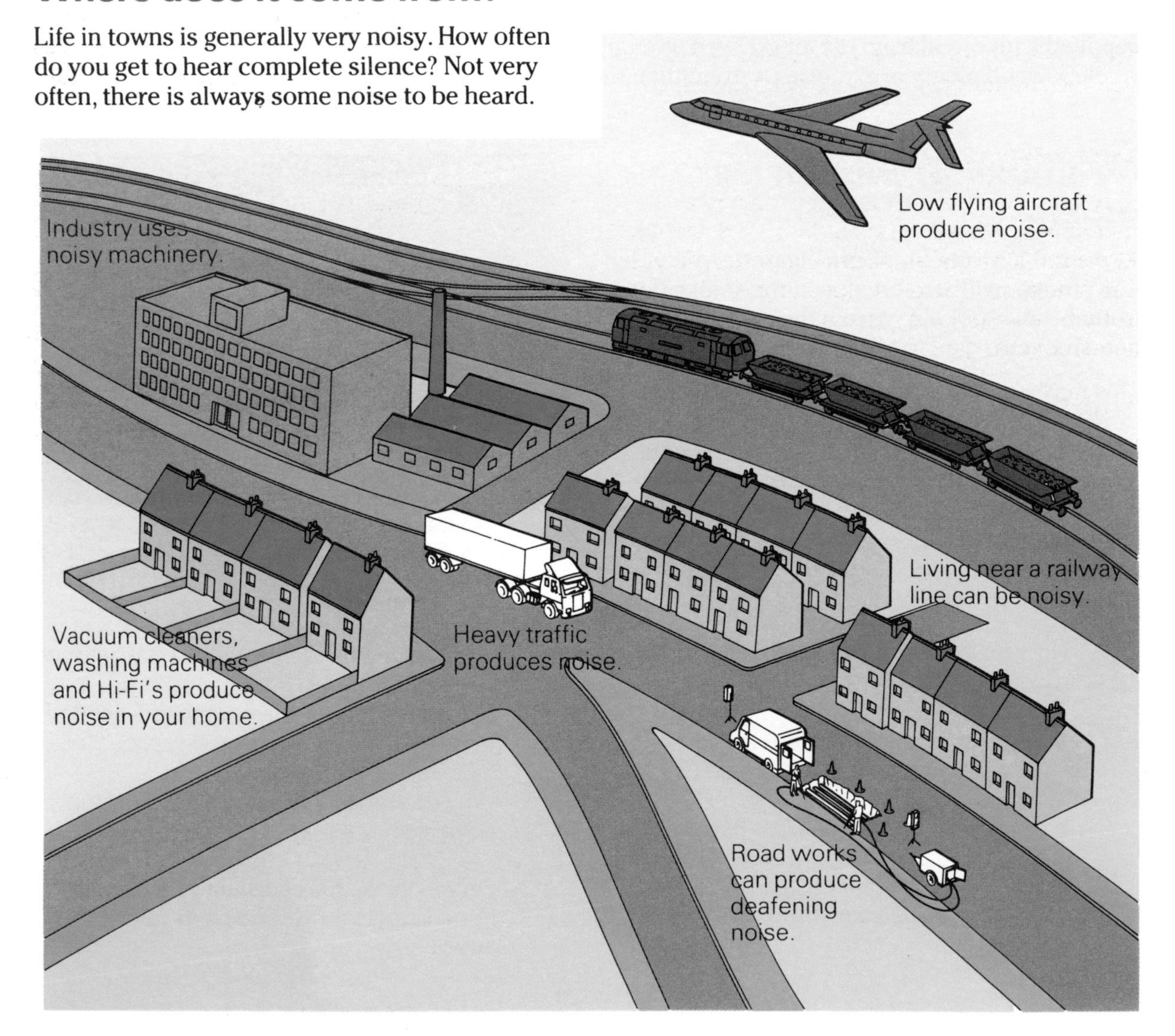

How loud is loud?

The amount of energy or the level of noise is measure in **decibels** (dB). It is measured by sound level meters. If one noise is twice as loud as another, the noise level increases by 10 dB.

So a vacuum cleaner making a noise of 70 dB is twice as loud as a normal conversation. Likewise, 120 dB from a low jet is twice as loud as a 110 dB road drill. The sound of an explosion may even have enough energy to break glass.

Type of Noise	Noise Level (dB)
Conversation	60
Door slamming	80
Road drill	110
Whisper	20
Vacuum cleaner	70
Low flying jet	120

Over the top

Constant exposure to noise levels greater than 80 dB may cause permanent loss of hearing. This can mean you will not hear quiet sounds and somebody else's speech may sound confused.

Many people like to use personal stereos to listen to music. These can damage your hearing if they are too loud. If a person at an arm's length away has to shout before you can hear them, then the sound is too noisy for your own good.

Reduce the sound level or you'll 'reduce' your hearing!

Keeping it down

To reduce the level of noise, you have to use materials which will **absorb** the sound and not reflect it. This is called **sound-proofing**. In your home, there are materials such as carpets and curtains which will act as sound absorbers. In factories, noise levels have been reduced by supplying ear mufflers and using sound-absorbing screens.

By changing the material from which a machine is made you can also reduce the noise it makes. For example, using bronze gear wheels to replace much noisier steel ones in machinery. By using plastic parts even road-work drills can be made quieter!

Ear mufflers protect this worker from the damaging noise of this tree stripper.

1 Name two possible sources of noise pollution.

2 Look at the table of noise levels.
- **a** What units are used to measure noise levels?
- **b** Which noise is the loudest?
- **c** Which noise is twice as loud as a vacuum cleaner?

3
- **a** Name two jobs where workers could be exposed to more than 80dB of noise.
- **b** What harm could happen to these workers?

4
- **a** Explain how screens can reduce noise levels in factories.
- **b** What type of material could they be made from?

5 What precautions should be taken to protect the hearing of a jet pilot?

2.13 What a waste!

A load of rubbish

In Britain, about 25 million tonnes of rubbish are produced each year. Apart from the huge mass involved, the volume of the rubbish presents problems too. And some of it is dangerous and poisonous as well!

This makes the problem of disposing of it very difficult. Local councils are responsible for waste disposal and have to find answers to these problems.

Contents of waste	1930	1980
Dust/cinders	55%	10%
Vegetables	10%	25%
Paper	15%	30%
Metal	5%	10%
Glass	2%	8%
Plastic	0%	10%
Other	13%	4%

There is much more rubbish nowadays and the type of rubbish has changed too.

Landfilling

Rubbish is usually buried and left to rot under large areas of land. At first, the rotting rubbish is smelly and can produce methane, a gas which burns. Once the rubbish has rotted, the land slowly settles back to its previous level. After some years, the land can then be used again.

But many plastics do not rot away. Land that is filled with this kind of rubbish does not settle very quickly. To help overcome this problem, the rubbish is sometimes **shredded** to small bits. These are **compressed** into bails and then buried. The land used to bury bails settles quickly and can be used again after only a few years.

Large holes have to be dug to bury everybody's rubbish.

Making use of rubbish

Some councils have developed more useful ways of disposing of rubbish. It is **incinerated** (burned) in a furnace at high temperature. In some places, the energy released is used to heat homes, greenhouses or even to drive electrical generators.

The remaining materials are **separated** into ash and metal. Councils can then sell this metal to help repay the cost of the incinerator and other equipment. The ash is easily buried and does not take up the same amount of space as the original rubbish.

Getting something out of nothing – the Altrincham incinerator.

Don't make a mess!

Most councils solve their rubbish problems by landfilling. This often seems the easiest and cheapest way, but it can pollute the environment.

Animal and plant habitats are destroyed when the ground is opened up for landfilling. Chemicals from the rubbish can seap out into streams and ponds for many years afterwards. During the landfilling, people cannot make any use of the area.

Many councils are aware of these problems and the dumping of dangerous rubbish is strictly controlled.

Careful disposal of rubbish means land fill sites can be safe to use in the future.

High-cost rubbish

The cost of new methods such as shredding, incinerating and separating is very high. The machines used to do the work may cost the community a lot of money. But this should be balanced against the cost to the environment. If rubbish is properly treated before disposal, the environment will not be in danger.

In some places, landfill sites are used for building new houses. If the landfilling was carried out safely, there should be no problems. But in the past, there have been places where the land had not been allowed to settle properly. Before long, the new houses needed expensive repairs – to mend the damage caused by bad landfillings.

Method	Cost	Speed	Risks
Landfill	£3 (per tonne)	very slow	seepage, landslips
Shredding and landfill	£7 (per tonne)	slow	seepage
Incinerating and separating	£15 (per tonne)	very fast	pollution from waste gases

1
- **a** Which type of rubbish was not present in 1930?
- **b** Which type has decreased the most since 1930? Give a reason to explain this decline.

2 What is the most common way that councils get rid of rubbish?

3
- **a** What sort of rubbish needs shredding?
- **b** How does shredding rubbish help in its disposal?

4 What are the advantages and disadvantages of using an incinerator to get rid of waste?

5 Would you mind living in a house built on an old landfill site? Explain the reasons for your answer.

2.14 Products that don't pollute

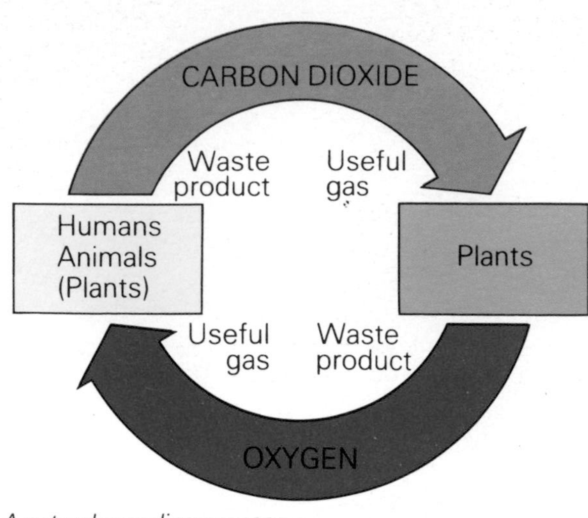

A natural recycling process

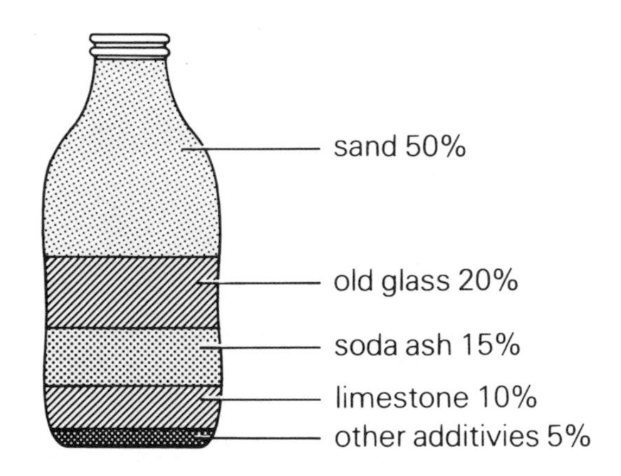

Turning sand into new glass needs a lot of heat. Recycling old glass uses less heat.

Natural cycles

Your body produces waste carbon dioxide gas which you get rid of by breathing out. This gas is used by plants to help them live. Similarly, oxygen is produced as a waste gas by plants – but you need it to stay alive. The use and re-use of different materials between animals and plants is a **natural cycle** – a way of sharing the resources of the environment.

Metal cans, glass bottles and paper often end up as waste. A lot of energy and resources are needed to make these items and often they are just thrown away as rubbish. Not only is there waste, but often pollution too! More energy and more resources have to be used *just to replace* these items. No cycle here – just waste, waste and still more waste!

Recycling our rubbish

About a third of all our rubbish is made up of packaging that comes from metal, paper and glass. If all this packaging was not put in the dustbin, £50 000 000 a year could be cut from the national waste disposal bill.

If these materials could be **recycled** (used again), then even more money could be saved by reducing the amount of energy and resources needed to make replacement packaging.

During the Second World War, many things were in short supply – and recycling was commonplace. After the War, people wanted more convenience. There was plenty of resources and so rubbish was just thrown away. But nowadays things are being recycled – to save money and to protect the environment.

'Gotta lotta bottles'

Glass bottle manufacturers have developed a scheme where used bottles can be returned. People can throw their used bottles into a **bottle bank**. This glass is then used to make new bottles at a lower cost. All the manufacturers have to do is to melt down the old glass. The use of old glass to make new glass saves energy, helps to keep down the cost of new bottles. The local council usually sets up the bottle bank and gets paid by the manufacturers for the glass collected.

The 'cash a can' scheme

About 100 000 tonnes of expensive aluminium metal is thrown away each year. A recycling scheme has been set up to reduce this cost. Many schools make money out of this scheme by collecting old cans or 'ring tops'. This metal is then sold back to the manufacturers and the money used to help the school.

Iron and steel are often recycled in a similar way. The only problem is that large amounts of iron are difficult to collect and return. Some councils use special separating machines to collect these metals from their waste.

Making aluminium needs a lot of electricity. Recycling it helps to reduce costs – and reduces acid rain too!

Paper money

Nearly one million tonnes of paper are used each year. Making paper not only uses up important resources (trees), but also needs a lot of energy and releases a lot of pollution. Many organisations collect paper to raise money. The paper is then sold back to manufacturers, where it is recycled. The waste paper is then simply mixed with water, made into **pulp** and cleaned. After that it is ready for re-use.

Plastic pollution

Plastic materials are used more and more instead of paper packaging. But they do not rot away as easily as paper. Some new plastic materials are **biodegradable** – this means they will be decomposed by natural factors present in the environment.

Some biodegradable plastics will decompose when attacked by small organisms such as fungi (mould). Other biodegradable plastics are weakened by prolonged sunlight before breaking up into small pieces. These pieces are then completely decomposed by the action of fungi, small insects and other organisms involved in the rotting process.

Tree-saver! Look for this emblem on the back of recycled paper.

1 Give an example of how materials can be used and re-used in a natural cycle.

2 Give three examples of materials which can be recycled.

3 Explain in detail one scheme which recycles materials.

4 More and more plastic is used instead of paper
- **a** Explain carefully why this may be a problem
- **b** How it can be overcome?

5 What are the advantages and disadvantages recycling materials like metal, glass or paper?

2.15 The problem with pollution

There is a price to pay for our standard of living. Electricity is very convenient but power stations can produce acid rain. Cheap food is nice but the fertilisers and pesticides which are used to grow it slowly poison the environment.

Sometimes it is difficult to please everyone. Look at the problems facing the people who live on this island...

The harbour employs 300 people. Most people fish but some act as dockers for oil refinery ships.

The island has 2 800 inhabitants. Many are employed in hotels and restaurants for the tourist trade.

The "old" coal power station must be changed. A new oil or nuclear power station will be built for free. If the people want wind power, they have to pay all the costs – £500 each.

Tourists leave litter on the island's beaches.

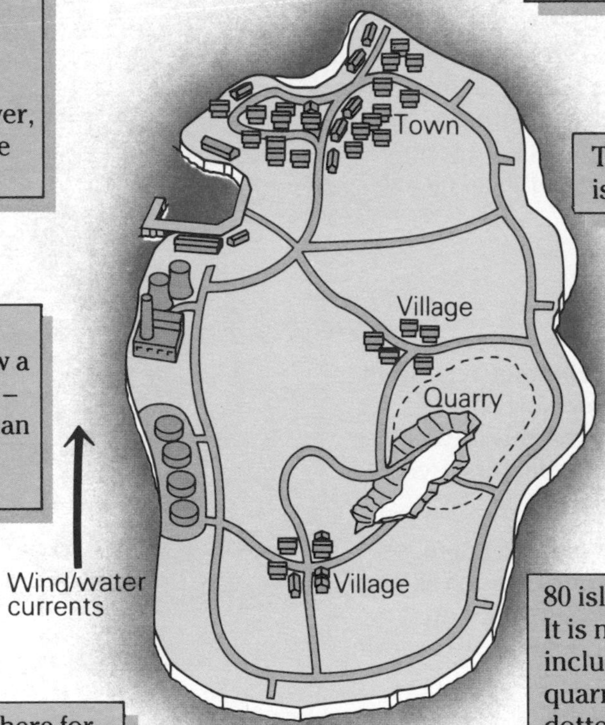

The island has been asked to agree to allow a nuclear power station – so that the mainland can get the electricity without the risk!

80 islanders work at the quarry. It is noisy and dusty. Plans include an expansion of the quarry, to take it up to the dotted line. The quarry will be exhausted in six years.

The refinery has been here for 15 years. There have only been two spillages, one on the beach and the other into a river.

Villagers used to farm sheep but oil, from a pipe at the refinery, polluted the river and ruined their grazing land.

More information about the island is on the next page. Find out more about the island and what is happening there and then try to answer some of the questions.

THE ISLAND NEWS

Beach disaster

Another oil slick threatens to ruin the holiday trade for the coming year. Hotelier Emma Lowe said 'Last year the beaches were a mess and next year we will lose even more business.' Local Councillor Anita Forster said that it was time that the refinery paid for its mistakes.

In an interview with the manager of the refinery he said that little could be done to stop small amounts of oil from seeping out. They would be mounting an inquiry into the matter as soon as possible. Until then, some new methods of stopping the oil from spreading were being tried.

Local M.P. and councillors meet

Charles Hutton the newly elected Member of Parliament, will meet councillors to discuss the island's future. Islanders are asked to attend the meeting tonight to voice their concern over a wide range of issues. In a recent TV discussion, the M.P. indicated his worry that tourism on the islands had taken a sharp decline since more and more industries had been set up on the islands.

Farm finishes

Another farm in the hills has closed as more people leave the island for work on the mainland. Some people may be able to get jobs with the Quarry Company if it is allowed to expand the scale of its mining works.

Record fishing catch

Young fisherman Mike and his wife Bethan have hauled in a record catch. Their new boat run by their family went out into deep waters to get their fish. Young fisherwoman, Bethan said that it was very hard work but they had to leave the polluted waters close to the island to find the fish. In recent years the size of their catch has dropped to less than half it was 5 years ago.

1. Identify as many as you can of the sources of pollution which are already on the island.
2. Which pollution risks may be added to these in the future?
3. Which industries benefit the community?
4. If you attend the meeting with the M.P. and the councillors, what would you have to say?

3.1 Healthy organisms

Alive or dead!

Anything which is alive is called an **organism** – but how do you know if something is alive? These are the main features of life:

Living the Good Life

There is more to being alive than just the seven features given above. An organism can get all it needs to stay alive but it may not be healthy. What do you think the word **healthy** means? Look at the people in these photographs. Do you think they are healthy?

Can everyone expect to be healthy?

Your health can depend on where in the world you live. A woman in rural India will probably live to be about 45-50 years. A woman in a European town will probably live to be over 70 years old. The picture below tells you more.

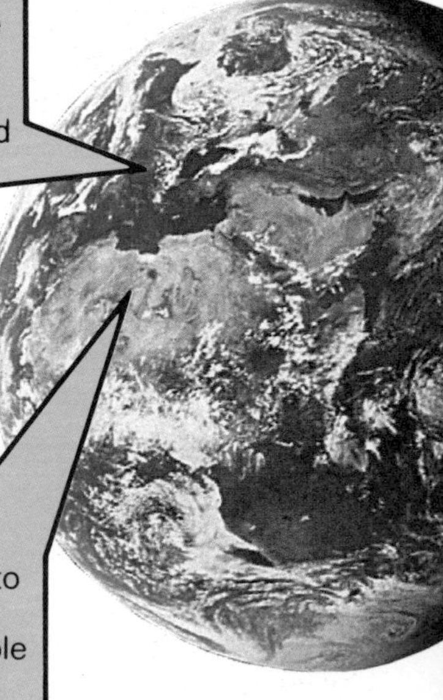

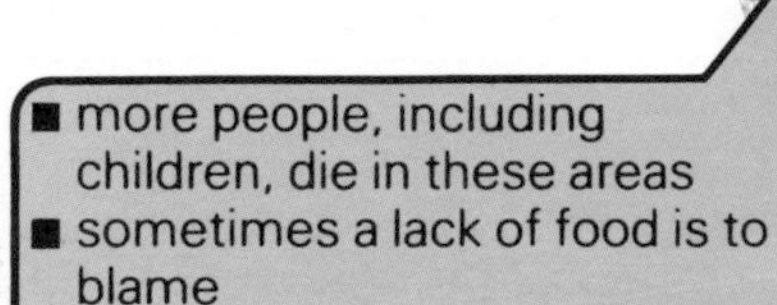

What happens if you stay healthy?

Your body is a complicated organism. It is made up of millions of tiny parts called **cells.** These cells work together to keep you alive. Your cells need certain conditions to keep them working properly. When you are 'healthy', your food and life style provide these conditions. If you stay healthy your body keeps working properly. This helps you to avoid illness and injury – even if you do fall ill, you get better more quickly. Being healthy makes you feel less tired and often lets you think more clearly. You will probably live longer too!

How do people stay healthy?

Think of the healthiest man (or boy) and woman (or girl) you know. Think of all the reasons why you consider them to be healthy. You should be able to think of at least five reasons.

Now write out a list of these reasons – put the most important ones at the top of the list.

Posters are often used to advise people of how to stay healthy. Look at these posters – what things are featured in them?

Draw a poster which would tell people about the most important reasons in your list. Try to make it as attractive, interesting and informative as you can.

What about you?

You are alive – so you must show the seven main features of life. But what exactly happens when you eat, breathe, grow or move? Why do you have to get rid of waste? What use is it to be able to react to what happens around you? You can learn a lot about yourself in this module. Once you know how your body works, you will also know how to stay healthy too!

3.2 Using energy

In a race, a sprinter uses enough energy to boil one cup of water.

Why do you need energy?

You need energy to do *anything*. Energy is used to keep warm, to go for a walk and even to grow. You also need energy just to stay alive – even when you may believe you are doing nothing! The amount of energy you need depends on what you do.

Where do you get your energy?

You get your energy from the food you eat. This is because your food is made of special chemicals which have energy stored inside them. For example, there is a lot of chemical energy stored in foods such as pasta. Your body can release this chemical energy from the food. This energy is used for warmth, movement, growth, or even just staying alive.

If humans don't get enough of energy from their food, their health may suffer. A child would become very tired and would not grow much. An old person would feel cold all the time and could even die from getting too cold.

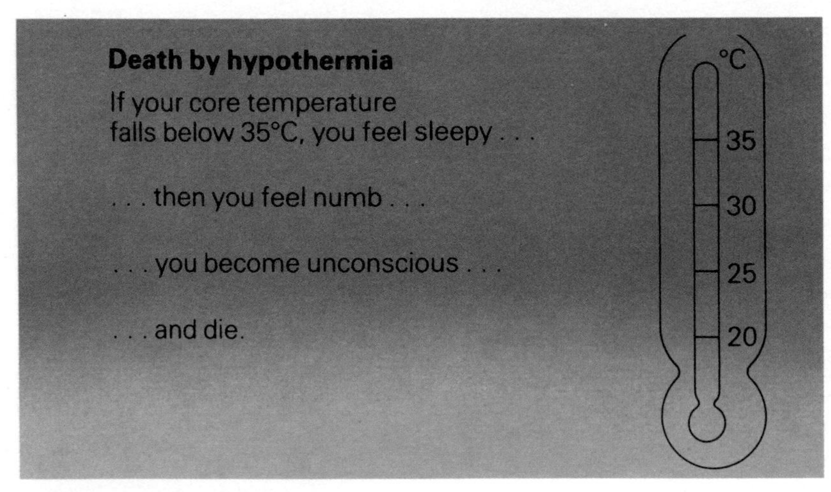

Energy for warmth

Some energy is needed to keep your main body organs warm. The temperature of these organs is called your body's **core temperature**. To keep these organs working properly, your body's **core temperature** must be kept above 35°C. But your body loses heat energy all the time. If too much heat is lost, your core temperature will fall below 35°C. When this happens, you are suffering from **hypothermia**.

Dying from the cold

As well as eating enough energy-rich food, warm clothes and hot drinks can help to prevent hypothermia. The heat from the drinks warms the body from the 'inside-out' and the clothes help to keep that heat in.

Old people may suffer from hypothermia in winter, but others can be at risk too. Far away from the shore, the sea can be freezing cold – especially in winter. If someone falls in, the ice-cold water soon soaks through their clothes. This can cause hypothermia in minutes!

The high cost of heating a home in winter can put old people at risk from hypothermia.

Energy for movement

Walking and running use energy – but so do all other movements; eating, talking, writing and breathing. Any movement needs energy to make it happen. The bigger the movement, the more energy is needed.

Most sports involve a lot of movement. These bar charts show you why you get hot when you are more active.

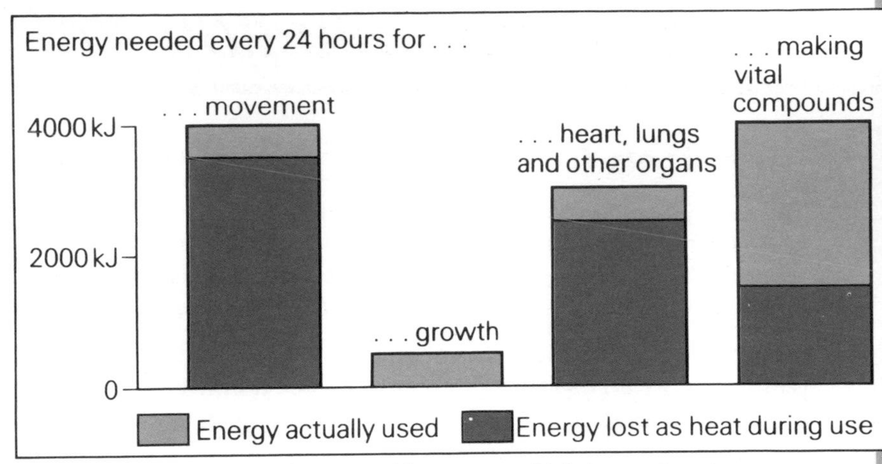

Although energy is used for many things, most of it is lost as heat!

Energy for growth

You have grown in size and weight since you were born. As you grow, you use energy to help make the cells of which your body is made. Even when you have stopped growing you still need to make new cells to replace old ones as they die off. You can lose 4 million skin cells in a lesson! They just die and eventually fall off. So you need energy all the time to make new cells.

Energy to stay alive

At night while you are fast asleep, your body is still active. Breathing, repairing cells, fighting infections – these all involve important chemical reactions which keep you alive. Energy is needed to keep these reactions going. These important reactions carry on non-stop, so you need a constant supply of energy from food to stay alive. The table shows you that even when you sleep you use as much energy as the lightbulb in your room.

Activity	Energy used in 1 minute (measured in kilojoules)
Lying still	5 kJ
Sitting	7 kJ
Walking	10 kJ
Dancing	30 kJ
Running	44 kJ
Swimming	46 kJ
Skiing	64 kJ cold water → hot water

1 Energy is used for all kinds of movements. Write down two other things energy is used for.

2 **a** Where does your energy come from?
b Why do marathon runners eat lots of pasta before a race?

3 Why do you get hot when you run around a lot?

4 **a** Who do you think would use up most energy, someone who is sleeping or someone who is running? Why?
b Who needs more energy *in total* – two people dancing or twelve people sleeping?

5 **a** What happens to your body when you suffer from hypothermia?
b Why do you think that old people are more likely to die from hypothermia than young people?

3.3 What are you eating?

Chemicals for dinner!

The food you eat gives you the energy and materials you need to keep your body working properly. Most food contain many different chemicals. Scientists divide these chemicals into five main groups: carbohydrates; fats; proteins; minerals; vitamins.

Different foods contain different amounts of these chemicals. This picture shows an example of the British diet. It may look like 'just food', but it contains the many chemicals you need to stay alive.

Carbohydrates

Carbohydrates are chemicals that are used by your body to *store* energy for a short time. They can *produce* energy quickly when you need it. Carbohydrate foods are usually cheap and filling foods, eg, bread, rice, pasta and flour.

Fats

Fats are chemicals that are used to *store* energy for a long time. Fats cannot release energy quickly. This means it takes a long time for your body to use up any stores of fats. They also help to *insulate* your body so that you do not lose a lot of heat. Butter, margarine, meat and most fried foods contain a lot of fat.

Proteins

Proteins are used to help your body *grow*. They are used to make new cells and to repair damaged ones. If you do not get enough protein in your food, growth stops. Your body then has to use proteins already inside your body. This means your muscles start to waste away. Proteins may be used to give energy, but only if you do not get enough carbohydrate or fat. Meat, eggs, cheese and nuts all contain proteins.

Minerals

Minerals are simple chemicals which are found in tiny amounts in your food. They are substances such as calcium, iron and iodine. These minerals are essential for cells to work efficiently. Iron is needed to make haemoglobin – a complex chemical which plays a vital role in the breathing process.

Vegetables contain minerals and vitamins.

Vital vitamins

Vitamins are usually complex chemicals, but they are given simple names: vitamin A, vitamin C, and so on. They are found in small amounts in most foods. Vitamins help to make cells work efficiently. You cannot store many vitamins or minerals in your body, so you need constant supplies of both.

	What is it needed for?	What happens if you don't get it?
Vitamin A	To keep your lung tissues healthy. For light-sensitive cells in the eye	Lung infections Blindness
Vitamin B	To help many of the chemical reactions in your body	Tiredness, weakness, paralysis, and a disease called Beri Beri
Vitamin D	To help make use of minerals to make bones	Softening of bones and deformity (rickets).

Mineral	What it is needed for?	What happens if you don't get any?
Iron	To make **haemoglobin** in red blood cells	Tiredness, sleepiness (Anaemia)
Iodine	To make certain hormones. These control reactions in your body.	Possible weight increase swelling of the neck (goitre)
Calcium	Growth of bones and teeth	Brittle bones, teeth decay quickly

Some of the vitamins and minerals you need to stay healthy.

Fibre

Some foods contain a very tough material called fibre. This is not used by your body as a food. But it is still very important because the fibre helps to *clean* your gut. It can even soak up dangerous chemicals. Brown bread, brown rice, salads, fruit and vegetables all have a lot of fibre.

Food for thought

Look at the table below. Can you find any foods which you normally eat? Do you think you get enough of each type of chemical? Do you get too much of some others?

Carbohydrates	Fats	Proteins	Fibre, Minerals and Vitamins
High content CHOC BISCUITS CRISPS Chips	CRISPS CHOC BISCUITS		RICE
RICE BAKED BEANS Spuds		CHOC BISCUITS CRISPS BAKED BEANS	BISCUITS CHOC
Low content Cheese	BAKED BEANS RICE	RICE	CRISPS BAKED BEANS

1 What are the five main things that you find in the foods you eat?

2 What foods would you eat to:
a give you energy?
b help you grow?
c help your body work efficiently?

3 What else is found in some foods that your body also needs?

4 Choose three foods from the table shown above that *together* would give you all the things your body needs from food.

3.4 Eating the right amount

A balanced diet

Are you on a diet? In fact your **diet** is the food you usually eat. When people say they are 'going on a diet' they mean a 'special diet'. This means that they change the amounts and the types of food they eat. On a **balanced diet** you only eat the foods that your body needs. If you get the balance wrong and eat too little or too much of one food your health will suffer. If you eat too much carbohydrate for example, your body will turn it into fat. This will then be stored around your stomach, hips and legs – and you will become overweight. The results can even be fatal!

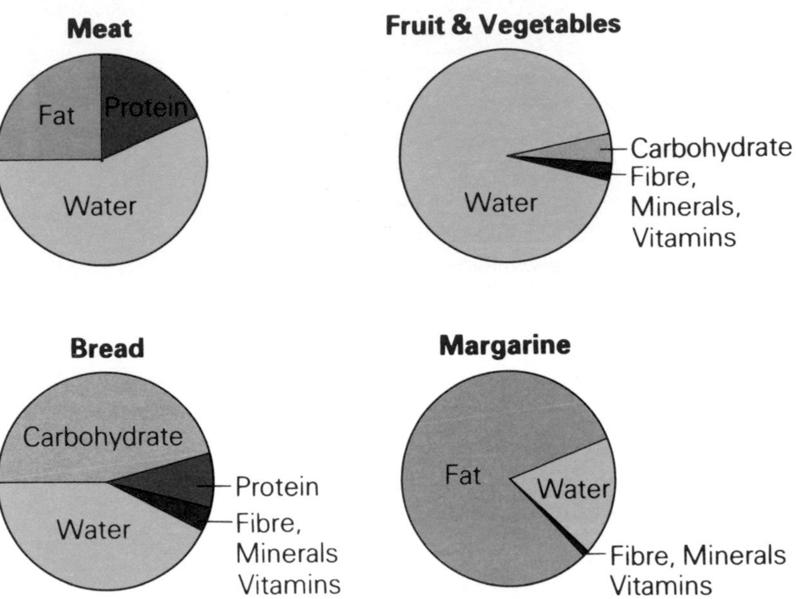

Did you know that you 'eat' a lot of water? These pie charts show you what your food is made of.

What do you need?

The *kind* of food you eat is important but so is the *amount* of food you eat. Each person needs different amounts of different foods. Why? It all depends on what age and sex they are and what they do. The charts show how people's needs can vary. They show the daily needs of fairly active people. Where do you fit in?

Teenagers need about the same amount of food as adults. This is because a lot of protein and energy is needed for growth.

Boys generally need more food than girls because boys grow to be bigger and heavier than girls. Although women usually need slightly less food than men, if a woman is pregnant she needs much more. After all, she is eating for two!

These bar charts show the masses of carbohydrate, fats and proteins needed by humans. The energy stored in each type of food is in red.

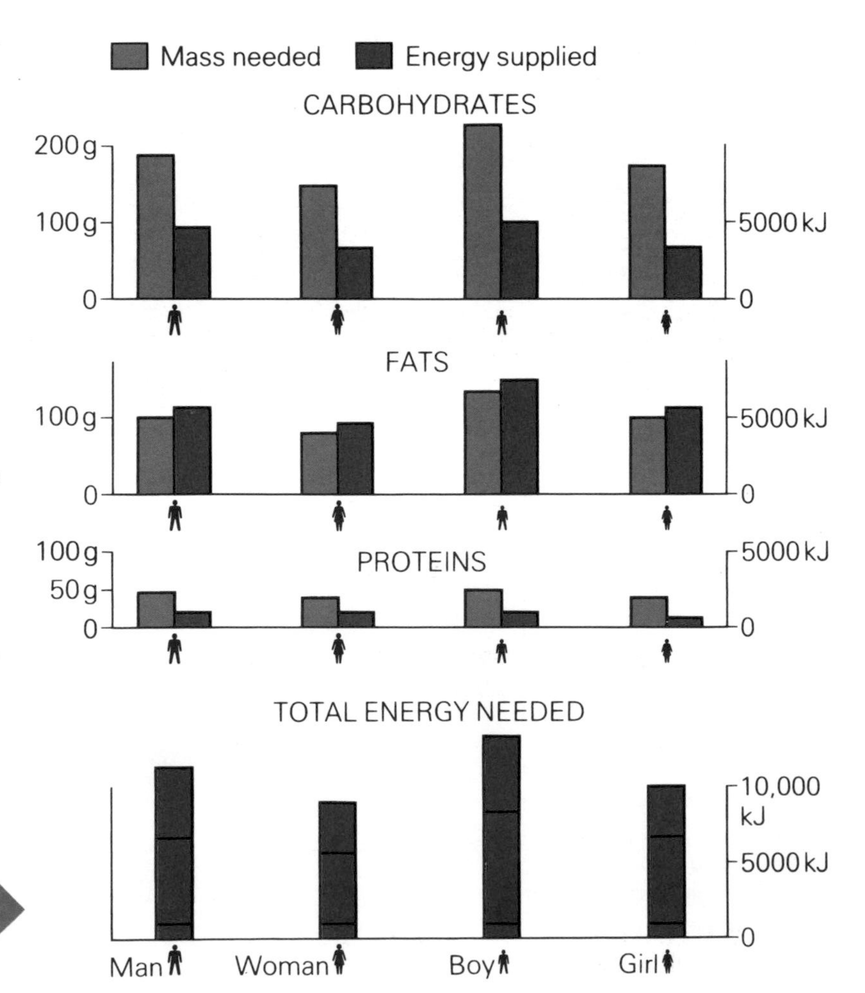

A poor diet

If the food you eat does not match your body's needs, then your diet is not balanced. In extreme cases a person may not get any carbohydrate or may take in too much fat. In each case their body will not get the nutrition needed in order to live a healthy life. Sometimes people suffer from **malnutrition**. This means the food in their usual diet does not match their food needs. Malnutrition can be caused by eating too much food as well as too little. The effects of malnutrition kill people, especially children. Over 40 million people die each year from a range of illnesses that are related to their diet.

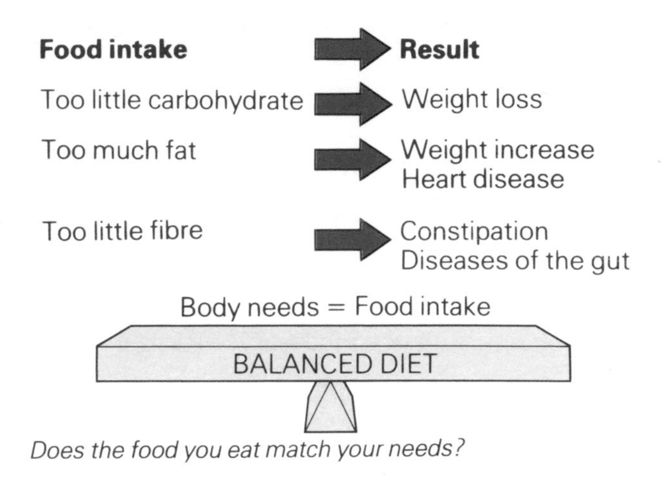

Does the food you eat match your needs?

What do you eat?

The tables below each picture show what these people eat in one average day. Two cases show malnutrition – one from eating too much, the other from eating too little. Which are they? Who has the most balanced diet?

Daily food intake provides:	
Energy	18600J
Carbohydrate	270g
Fat	180g
Protein	58g
Fibre	X

Daily food intake provides:	
Energy	9300J
Carbohydrate	196g
Fat	48g
Protein	28g
Fibre	✓✓

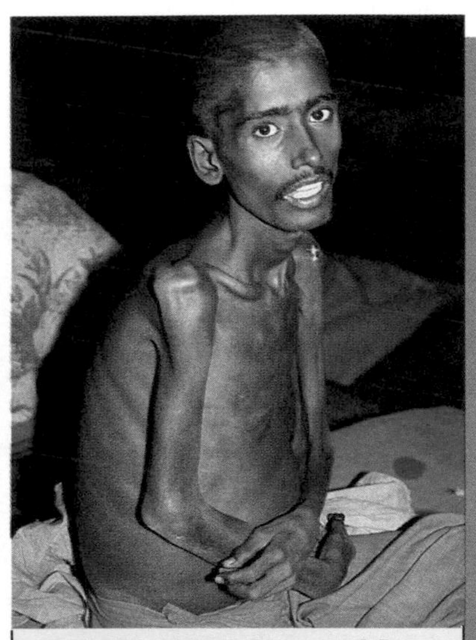

Daily food intake provides:	
Energy	2100J
Carbohydrate	45g
Fat	2g
Protein	6g
Fibre	✓

1 **a** What is meant by a diet?
b What is your usual diet?

2 What must be "balanced" in a balanced diet?

3 What are two effects of having a diet that is not balanced.

4 Explain what each person in the pictures above must do in order to balance their diet.

5 Which sort of foods should each person eat so that their diets will be balanced? (You may need to use the information on pages 66 and 67).

3.5 What happens when you eat?

Why do you eat?

You may think you eat just because you get hungry – but this isn't the only reason! The cells of your body need a constant supply of chemicals. These give the energy you need to live and grow. When you eat, you chew on food and swallow it. But what happens next? The chemicals in the food still have to get around your body.

Sometimes during hospital treatment, patients get the chemicals their bodies need without having to swallow anything. The chemicals that they need go directly from a drip feed into their bloodstream. These chemicals are then taken to all the cells in the body. These drips often contain glucose – a simple carbohydrate which will dissolve in blood. Glucose gives the patient the energy needed to get better.

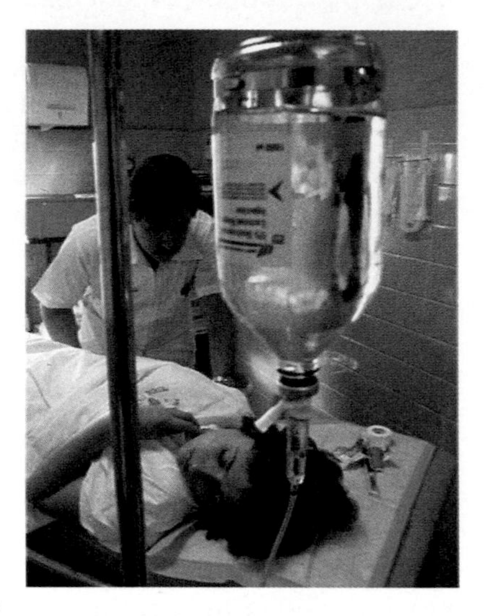

Glucose from this bottle flows down a drip feed, through a needle in the patient's arm and passes into their blood.

Getting food out of your gut

The chemicals in the food you eat must get from your gut into the **blood vessels** that surround your gut. Once the chemicals get into your blood, they can travel to every cell.

But your food is not usually made up of simple chemicals. For example, starch is a complex chemical – unlike glucose, it will not dissolve in your blood. This means that starch must be broken down into simple soluble chemicals. If this does not happen, the starch will stay in your gut.

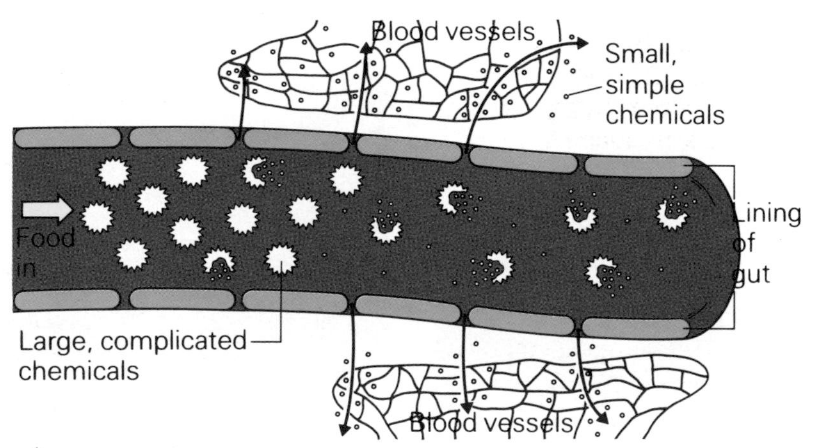

A simple chemical can pass out of your gut and into your bloodstream.

Breaking down your food

Your body has special substances that can help to break down the food you eat. These substances are called **enzymes**. When you eat some food, the enzymes in your gut mix with the complex chemicals in the food. Starch is an example of a complex carbohydrate. Through the action of enzymes, the starch you eat is *broken down* into a smaller carbohydrate called glucose. Glucose can then enter your blood by going through your gut.

Complex chemicals (such as starch) can be 'broken down' into simple chemicals (such as glucose) by enzymes.

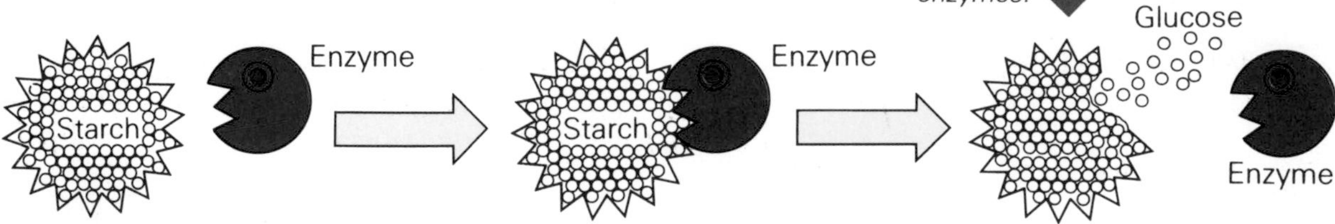

The stages of digestion

The whole process of breaking down your food into **soluble** substances is called **digestion**. In digestion, the food that you eat is changed into something that your body can use.

The process is a slow one which starts at the entrance of your gut – the **mouth**. Here your food is chewed. This breaks it into small pieces. Some enzymes and other body chemicals also mix with the food in the mouth.

Then the chewed food is swallowed and it is churned (mixed up) in the **stomach**. Other enzymes and chemicals are added here.

The mixture then moves into the **small intestines**. It is in this part of the gut that simple chemicals are made from your food. These can then escape into the bloodstream.

Digestion is a series of slow chemical reactions. This is the most important part of digestion – changing your food into simple chemicals.

Some chemicals from food (such as fibre) are not digested. The fibre in the waste helps to clean the gut and to absorb some chemicals that would harm you. Finally the waste collects at the **rectum** where it is ejected.

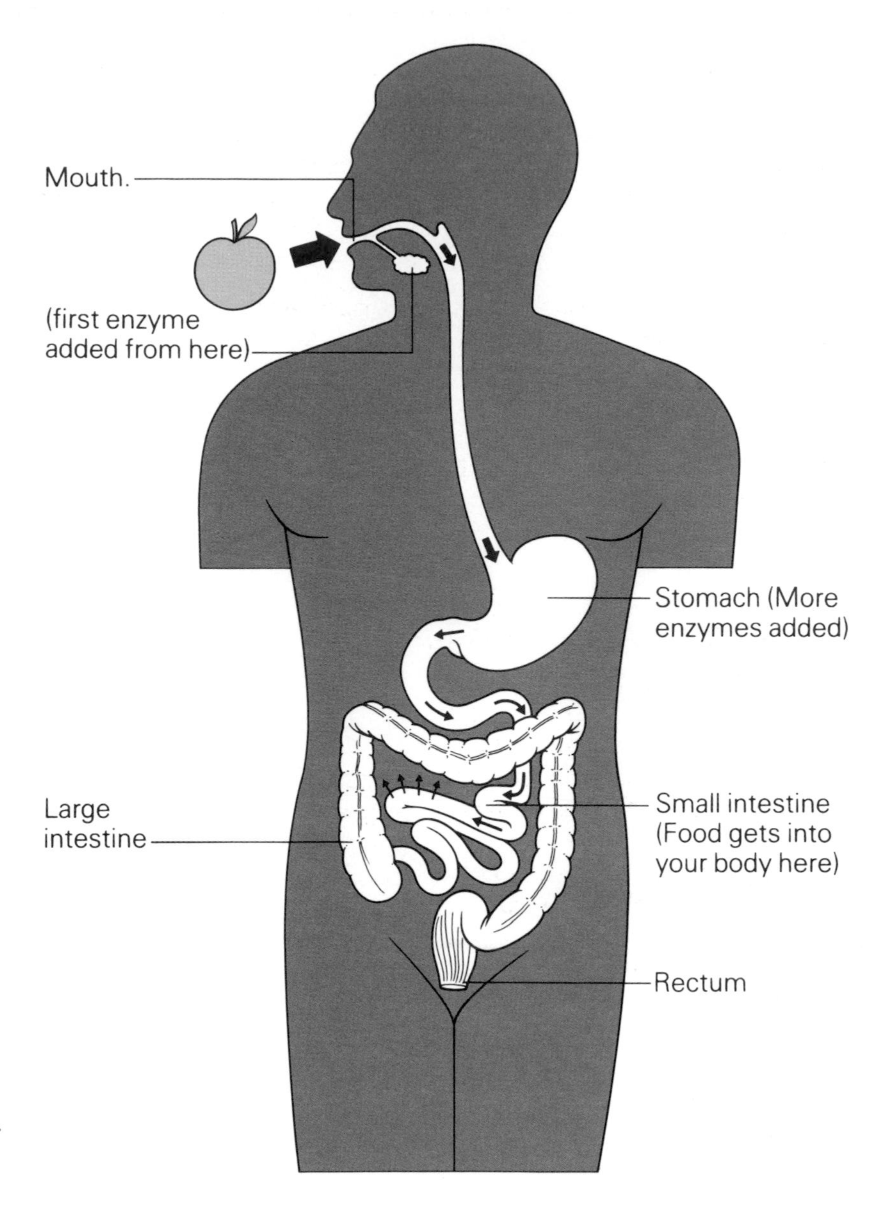

1 What is the difference between starch and glucose?

2 Are starch and glucose similar in any way?

3 What is an enzyme and what does it do?

4 Why does chewing help digestion?

5 Why are people who eat lots of foods which contain fibre less likely to suffer from problems with their gut?

3.6 Every breath you take

Glucose is used up by your body to give you energy.

'Finding' the energy

When you start to run you may be relaxed but soon you may be red-faced, gasping for air and tired! In very long races 'drinks stations' are provided. Glucose drinks help to give the runner more energy. But the runner also needs oxygen from the air.

Oxygen and glucose **react** together to release **energy.** But a lot of this energy is released as heat. This why you feel hot after running. **Carbon dioxide** gas and **water** are also produced when oxygen and glucose react together. This process is called **respiration.** Here is an equation (in words) which describes the respiration reaction:

Glucose + Oxygen → **Energy** + Carbon dioxide + water

Breathing – why bother?

How do you get the oxygen into your body? Your **lungs** do this important job. How do you get the carbon dioxide gas out of your body? Again you use your lungs. They **exchange** oxygen in the air for carbon dioxide in your blood. The water formed by respiration also needs to be removed. Can you think of any ways that water is lost? (You can find out more about water loss on page 78 and 79).

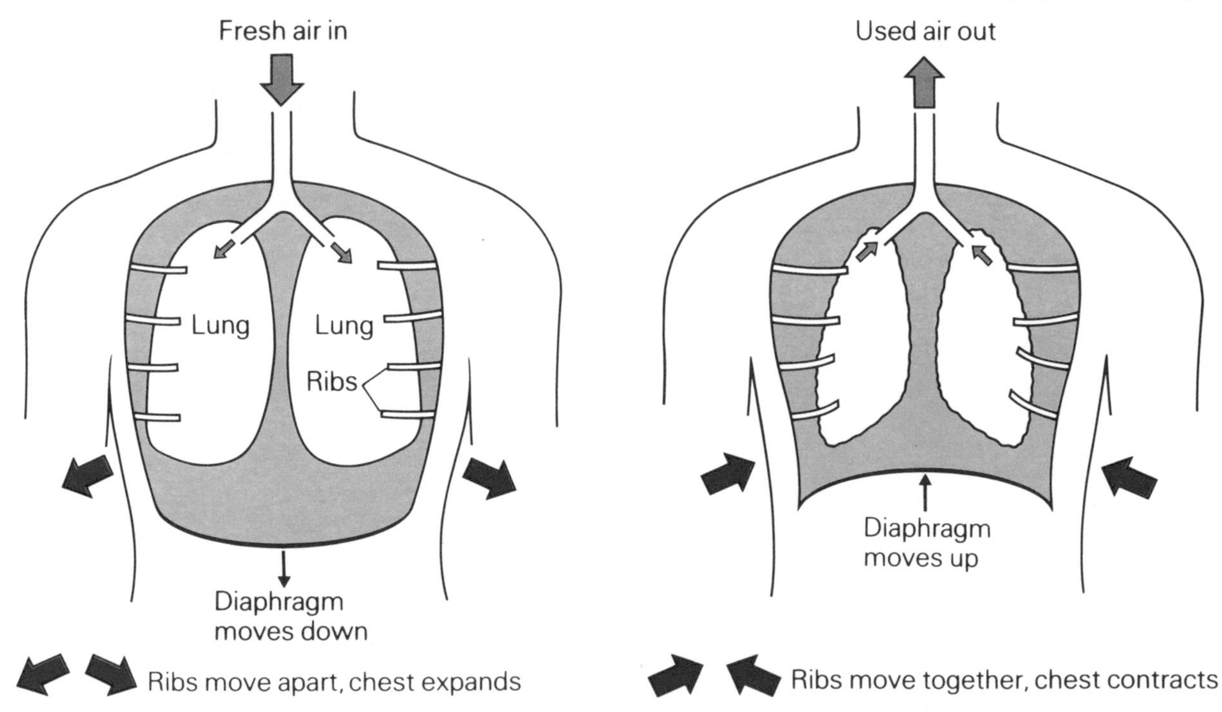

Your chest and diaphragm muscles suck fresh air into your lungs and push used air out.

Your lungs–the great exchange

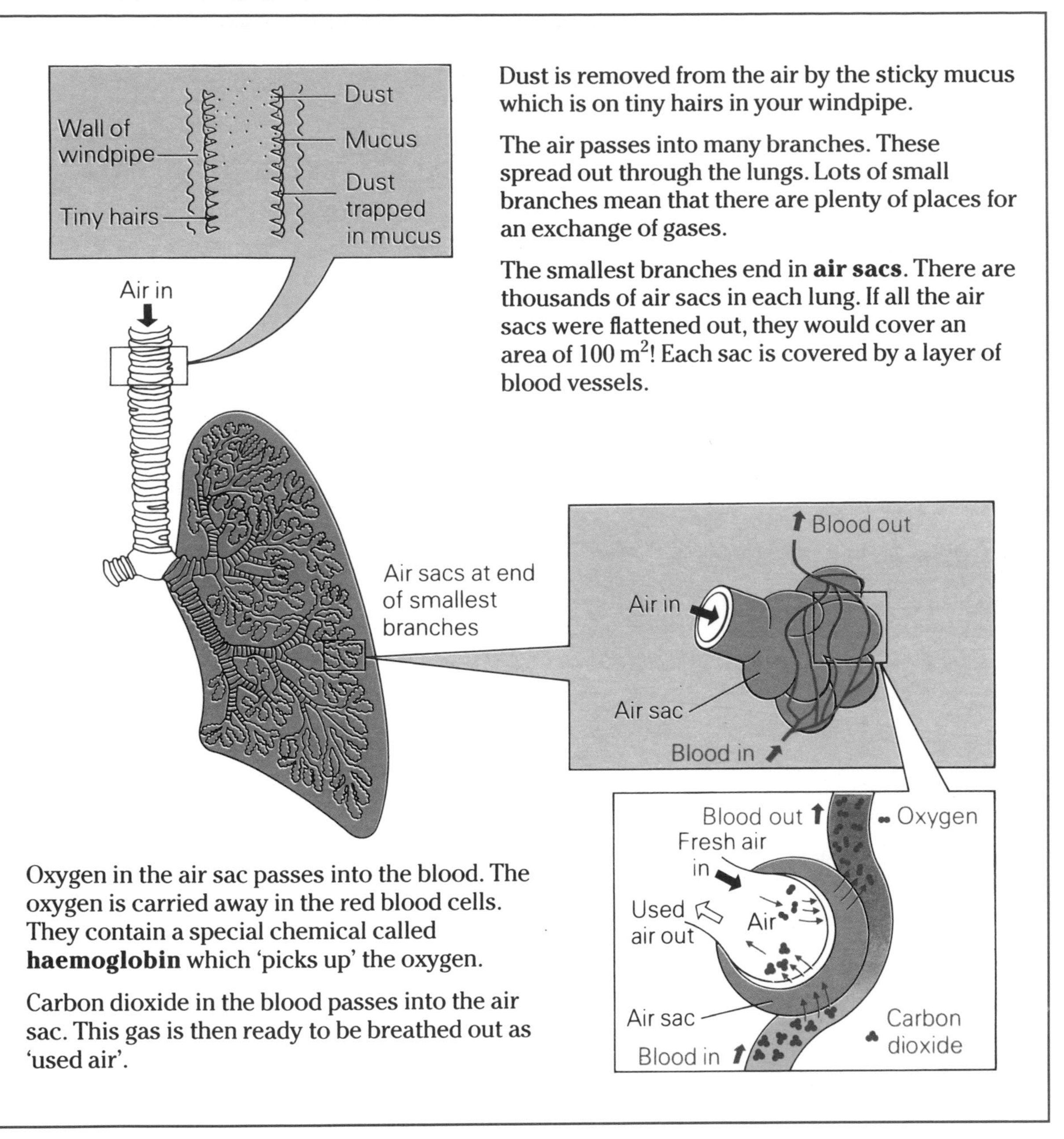

Dust is removed from the air by the sticky mucus which is on tiny hairs in your windpipe.

The air passes into many branches. These spread out through the lungs. Lots of small branches mean that there are plenty of places for an exchange of gases.

The smallest branches end in **air sacs**. There are thousands of air sacs in each lung. If all the air sacs were flattened out, they would cover an area of 100 m^2! Each sac is covered by a layer of blood vessels.

Oxygen in the air sac passes into the blood. The oxygen is carried away in the red blood cells. They contain a special chemical called **haemoglobin** which 'picks up' the oxygen.

Carbon dioxide in the blood passes into the air sac. This gas is then ready to be breathed out as 'used air'.

1
- **a** What two things do you need for respiration?
- **b** What three things do you get from respiration?

2
- **a** Your lungs do two important jobs. What are they?
- **b** What causes air to go in and out of your lungs?

3 How many desk tops would your lungs cover if they were folded out flat?

4 Marathon runners can get glucose from a 'drinks station'. Where do you get your glucose from? (You may need to look at page 70)

5
- **a** Where do you get oxygen from?
- **b** Why is oxygen important?
- **c** Explain why it is dangerous to put a plastic bag over your face.

3.7 Blood – supplying your needs

What does your blood do?

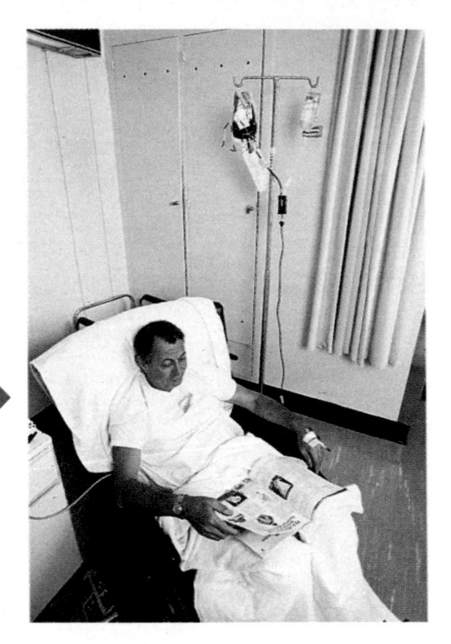

In many simple living organisms their body is made up of just one 'unit' called a cell. The chemicals that are needed for life are all inside this single cell. More complicated organisms have lots of cells. These organisms, such as humans, need to transport chemicals from one cell to another and their blood system does this job. An average 16 year old needs about 4 litres of blood to transport everything around the body.

If people have lost blood in an accident they can be given **blood transfusions.** This means giving them some 'new' blood to 'top up' their blood supply. Look at this picture. There is a small plastic bag hanging on the right which contains blood. The blood flows down the tube, then through a needle in the man's hand and into his body.

What is in your blood

Blood is a mixture of many different chemicals and cells. These all have different jobs to do. In the diagram you can see what is in your blood and what the different things do.

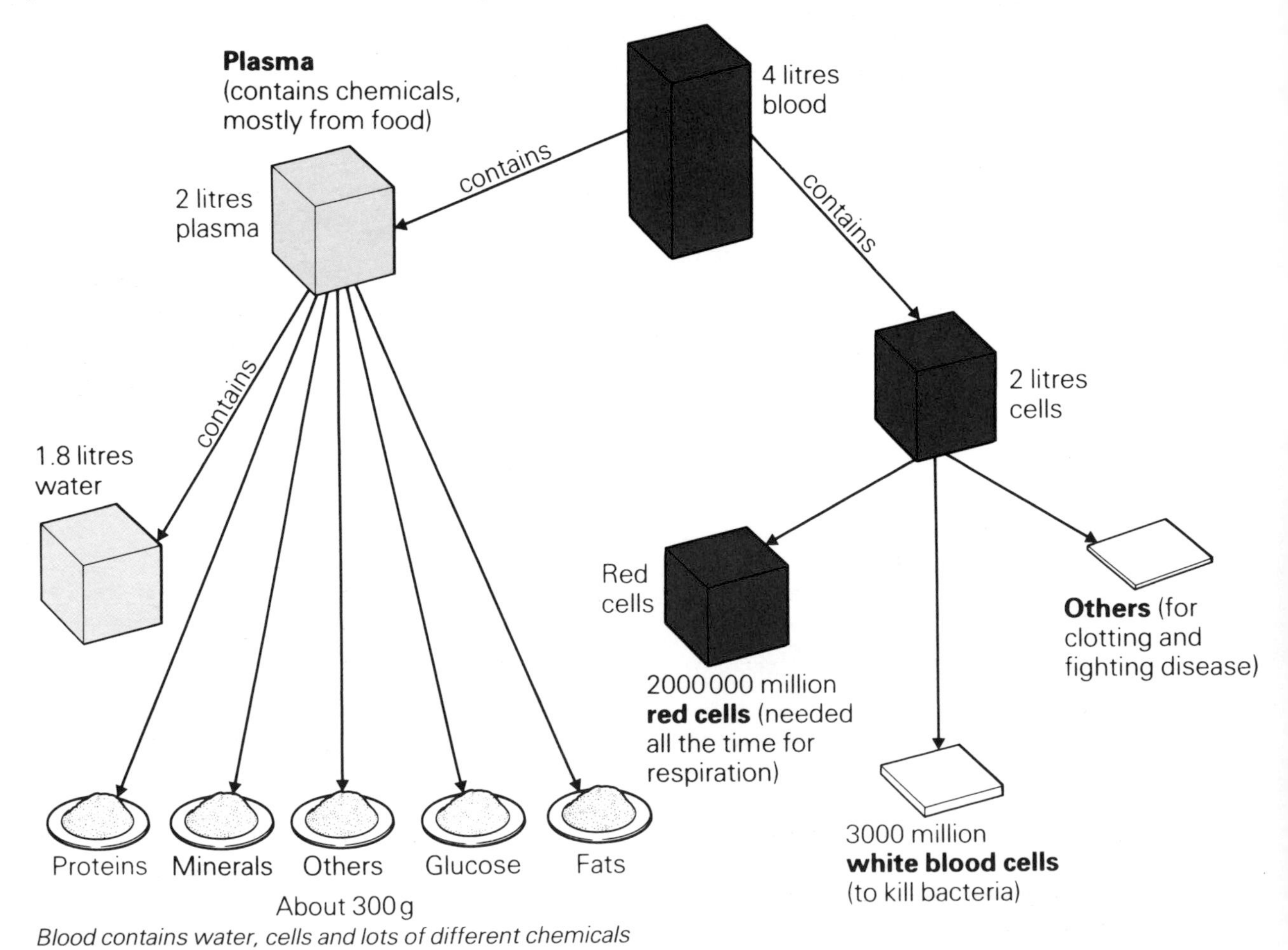

Blood contains water, cells and lots of different chemicals

Is everyone's blood the same?

When people bleed, it's always red, that's for sure! You have seen that blood contains many different chemicals and cells. But not all blood is exactly the same. There are four main types of human blood – O, A, B and AB – each of which contains slightly different chemicals.

The most common types are O and A; about 9 out of 10 people have either type of O or type A blood. These four types of blood are known as **blood groups**. If blood from the wrong group is used in a blood transfusion, the patient can become very ill. This is why careful checks are made before a blood transfusion.

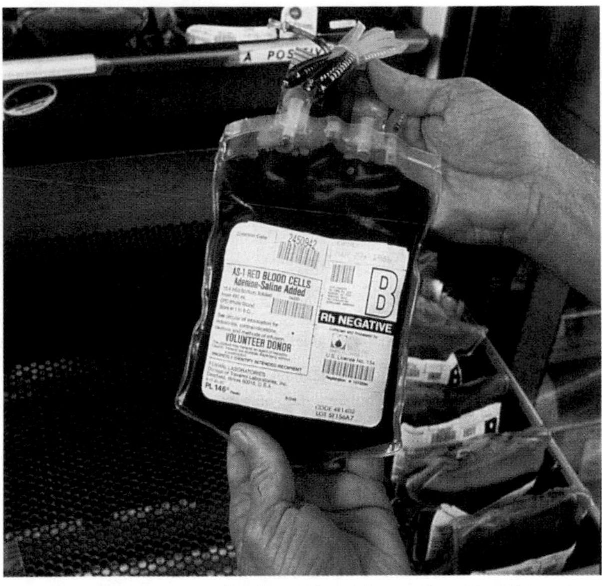

A blood 'bank' – notice that blood from the different groups is labelled clearly.

Bringing oxygen to every cell

All the cells in your body need oxygen to stay alive. It is your red blood cells that supply oxygen to all parts of your body. The red blood cells do this because they can join loosely with the oxygen. This is then carried around your body in the blood. The oxygen is then allowed to 'break off' to go to the cells which need it.

Carbon monoxide is a poisonous gas found in car fumes and cigarette smoke. It stops red blood cells from doing their job by blocking them up. The small amounts of carbon monoxide in cigarettes kill off muscle cells in your heart. This is why smoking weakens your heart. Large amounts of carbon monoxide can even kill you.

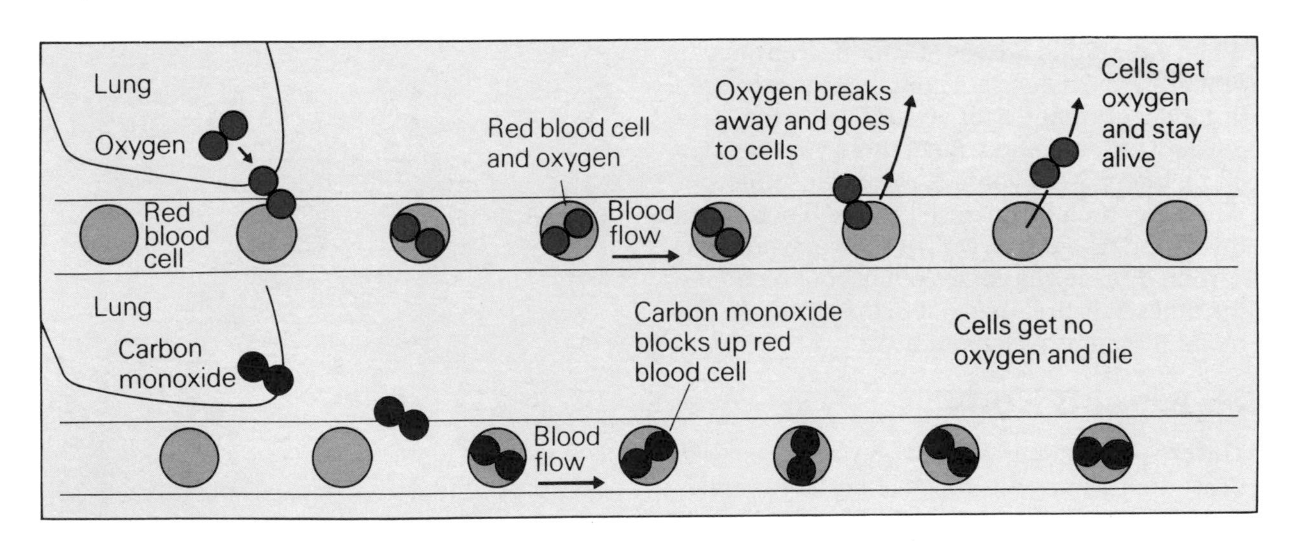

1 What happens in a blood transfusion?

2 **a** How much water does your blood contain?
b Where do the chemicals in your blood come from?

3 How does cigarette smoking damage your heart?

4 Why do you think your blood contains more red blood cells than white blood cells.

3.8 Your heart – a muscular pump

On the beat

Everything your body needs to stay alive can be found in your blood – oxygen, sugar, water and many other substances. A network of tubes called **blood vessels** pass through every part of your body. Blood is pumped along these tubes by your heart.

Your heart is made of **muscles**. These muscles form four hollow chambers. If the muscles of your heart tighten **(contract)**, the blood is squeezed out of the chambers. This contraction pumps the blood out of your heart and around the body. **Valves** stop the blood from moving the wrong way through the heart. The sound of your heartbeat is the noise of these valves opening and closing.

The muscles of your heart must contract in a special order if your blood is to be pumped efficiently.

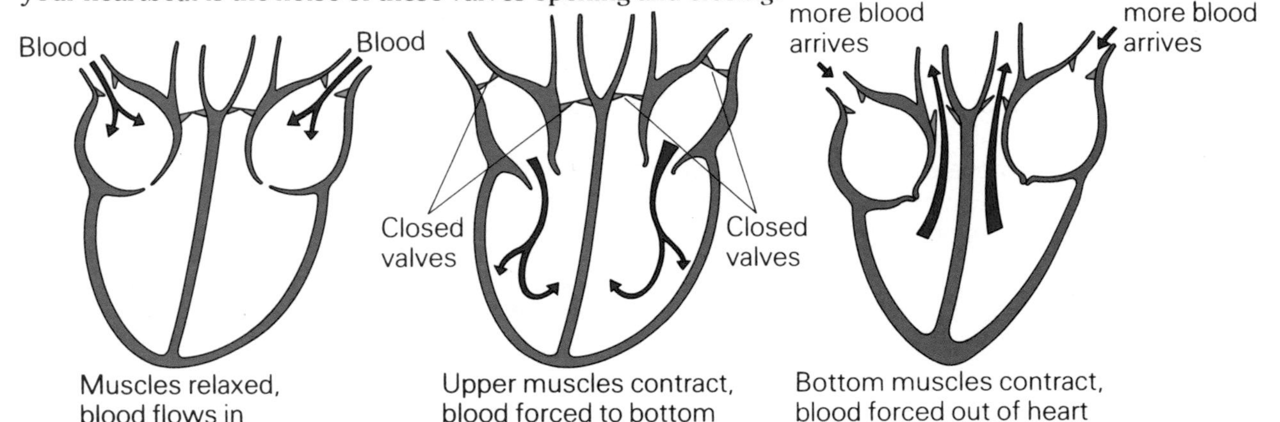

Heart and lungs

When your blood arrives at your heart from your other organs, it does not contain a lot of oxygen. But it does contain a lot of carbon dioxide – a product of respiration. Blood from your heart is first pumped from your heart to your **lungs**. When you breathe in 'fresh' air, red blood cells in this blood take up oxygen. At the same time, the carbon dioxide gas comes out of your blood in the lungs. It is breathed out in the 'used' air. The blood then returns to your heart.

Heart and Body

The blood now contains a supply of oxygen. Th heart then pumps this blood along large vessels called **arteries**. These take the blood to the major organs such as your brain, liver, kidneys and gut. These organs use up the oxygen in the blood during respiration.

The blood then returns from the organs to the heart. The blood does this by passing along blood vessels called **veins**. Once it is back in the heart, the blood is again pumped to the lungs, so that more oxygen can get into the blood.

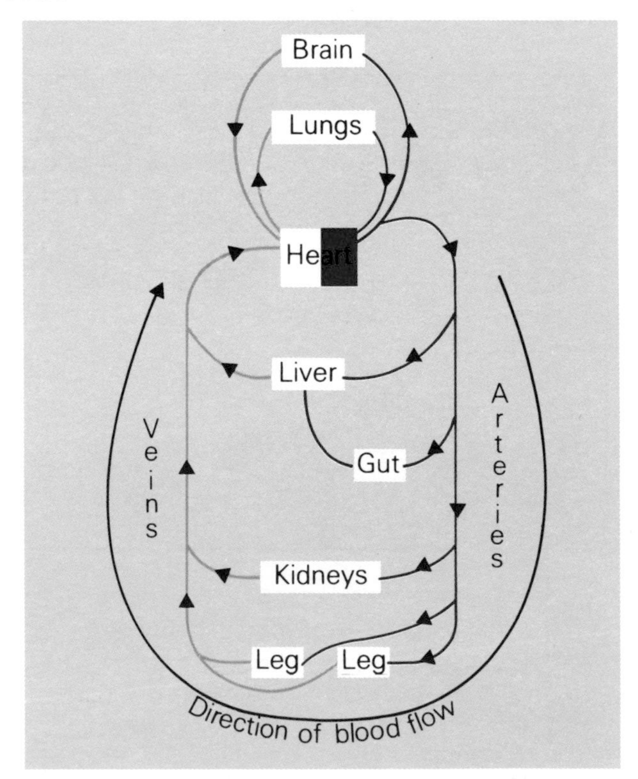

A simple view of how your blood flows around your body.

Supply lines

Some blood vessels carry a lot of blood but others can carry only a little. Their size and shape depends on where they are and what they have to do.

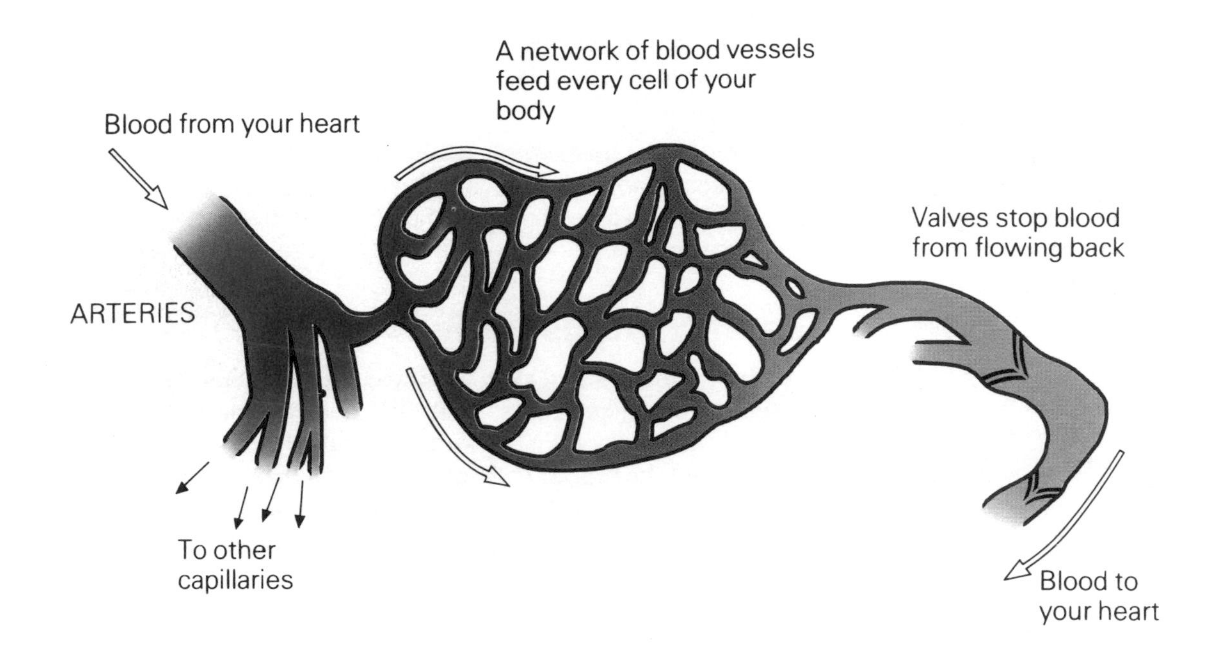

Large amounts of blood flow through the **arteries**. They bring blood to the major organs. The blood is under high pressure here so the walls of arteries are thick and strong.

Excess fat in your diet may cause blockages to occur in arteries. If there are any blockages, the results may be fatal.

Small amounts of blood flow through the **capillaries**. These narrow tubes can reach even the smallest cells in the body. This means vital chemicals can get directly from a capillary into most cells in the body.

There are thousands and thousands of capillaries in your body. They link your arteries to your veins.

Capillaries join up to form **veins**. These take blood back to the heart. A lot of blood flows through the veins, but the blood pressure is now low. This means the walls of veins do not need to be very thick.

Some small veins lie just under the skin. They can easily be seen as they have a faint blue colour.

1. What is the job of the heart?
2. Why is the blood pumped to your lungs first?
3. What are the names of the three types of blood vessels in the body?
4. Write down places in your body where you would find valves. What do they do?
5. Why do you need high blood pressure in the arteries?

3.9 Waste – life's leftovers

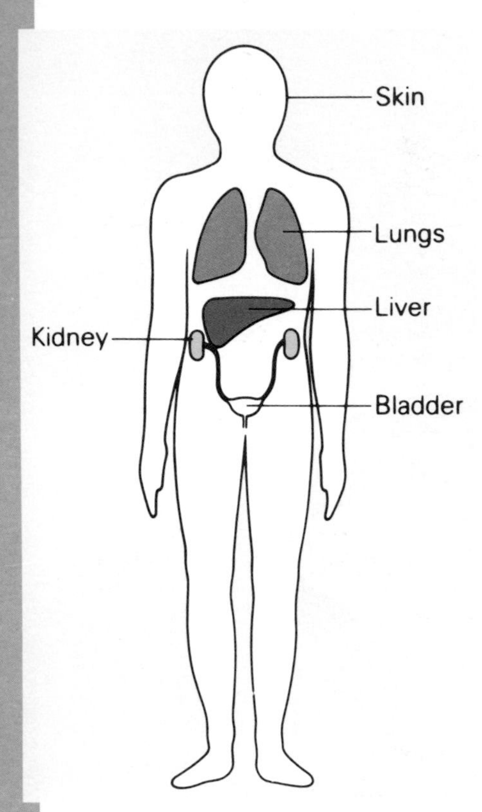

Each part of your body which is shown here helps to get rid of waste.

From 'junk' food to waste products

Your food contains many substances such as carbohydrates, fats and proteins. Your gut breaks them down so that they can be absorbed and used by your body. As your body uses these chemicals, waste products are made. These must be removed or they will harm you. The removal process is called **excretion**.

Carbon dioxide is an example of a waste product. It is formed during respiration. Your **lungs** remove the waste carbon dioxide. There are other waste products and these are dealt with by one of three organs.

Keeping things under control

Your **liver**, **kidneys** and **skin** are three other organs that control and remove waste products. Your liver controls many of the chemicals in your body. Your kidneys control most of the removal of waste from the blood. The heat produced by reactions inside your body is also a waste product. It is your skin that controls the ioss of waste heat from your body.

Life depends on the liver!

The blood in the vessels surrounding your gut contains many substances. This blood goes straight to your liver – your 'chemical factory'. In your liver, some of the proteins take part in chemical reactions. This produces a poisonous waste called **ammonia**. Your liver quickly changes the ammonia into a harmless chemical called **urea**. This 'safe waste' dissolves in the blood which flows out of your liver.

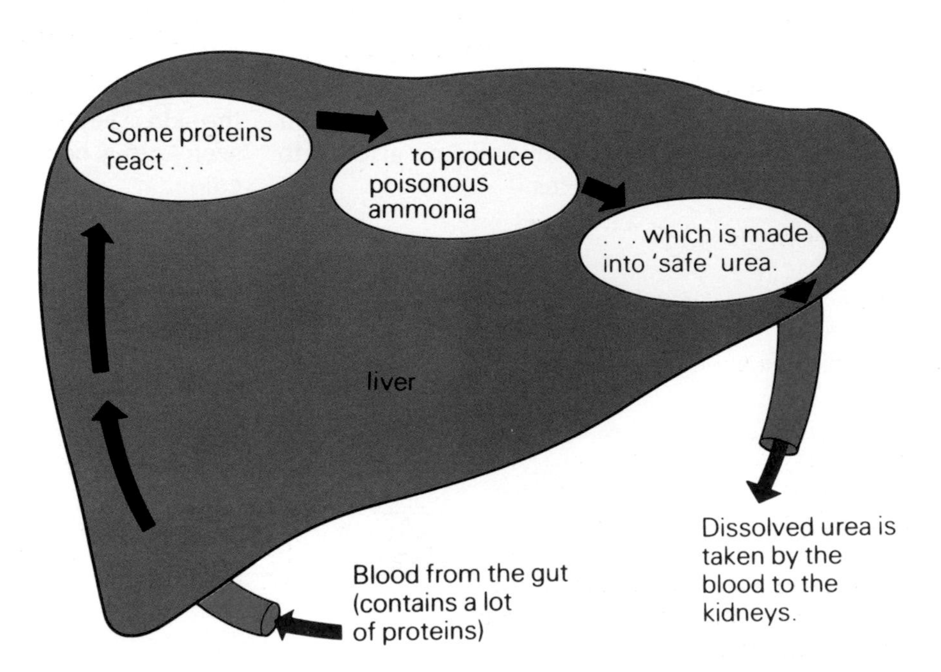

Filtering through

You have two kidneys in your body. They *filter* your blood and remove the wastes. Blood is pumped into very thin tubes in your kidneys. Urea, glucose, water and other simple chemicals are forced out of the blood. Useful chemicals are absorbed back into your blood. Any waste substances, such as urea, are not absorbed back. This waste is then diluted with water and lost from your body as **urine**.

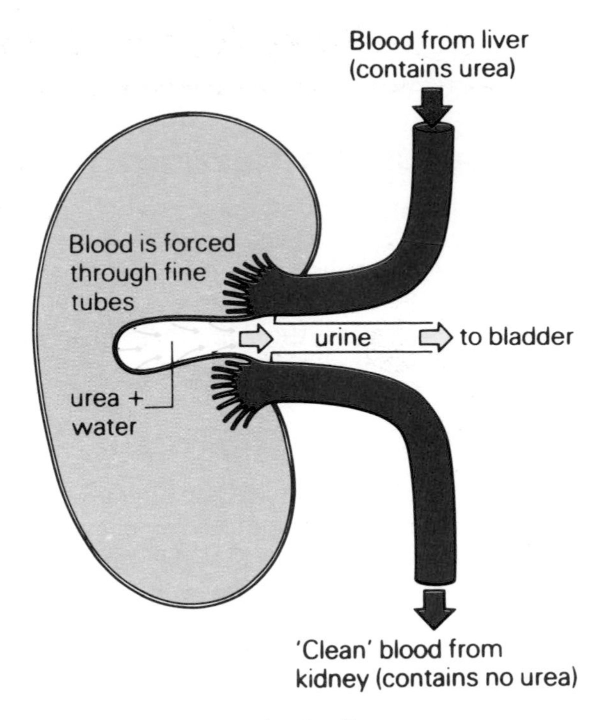

Your kidneys are very selective filters.

Skin deep

Your skin controls the loss of waste heat. It can do this by changing the flow of blood through your skin. When the blood is close to the surface, your skin looks red and heat passes from your blood through your skin. This heat is soon lost to the air and you cool down.

Some of the water and waste in your blood passes into sweat glands. These glands release the water and waste as sweat. The heat escaping through your skin causes the sweat to evaporate. This helps the heat to be lost even more quickly.

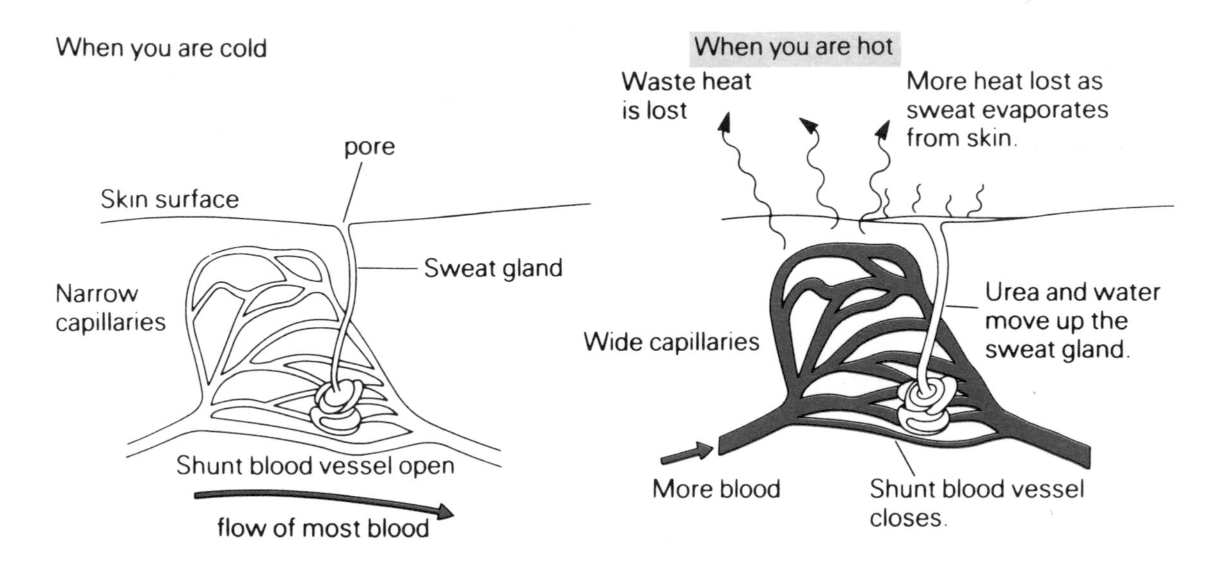

'When you're hot, you're hot!' – and your shunt vessels are closed!

1 Which organs of your body help remove waste?

2 **a** What happens to some of the proteins in your blood when they reach the liver?
b Why could this be dangerous?

3 **a** How does your liver deal with ammonia?
b What happens to the safe waste product made from ammonia?

4 **a** Explain how your kidneys remove waste from your body.
b What other waste products do you think could also be lost in your urine?

5 **a** Why is sweating important to your body?
b Why do you think you sweat more after running than walking?

3.10 Muscle & bone – a joint effort

Long-lasting protection....

Your bones are the hardest things in your body. Some bones have been found that are thousands of years old. Why do you need something so tough inside you? You have many vital organs in your body – if these get damaged you could die. Your chest has many bones just below the skin – these make up your rib-cage. This protects your heart and lungs from getting damaged.

....and a supporting role

Some organisms are supported by their surroundings. Certain jellyfish are completely supported by the water in which they float. You are surrounded only by air – this cannot support you. Your bones protect you but they also *support* your body. They can do this because they are linked together to form a **skeleton**. Without it you would collapse into a jumbled heap of organs and limbs!

On the move

Many of the bones in your skeleton are connected by special **joints**. These joints allow connected bones to move in certain directions. But what causes the bones to move? Every bone in your body has **muscles** fixed to it. You have to use chemical energy to make these muscles change length.

If a muscle becomes shorter, it will pull on the bone attached to it. The joint at the end of the bone twists round and the bone then moves in that direction.

Let's all pull together

Bones which have joints at the end usually have more than one muscle attached. These muscles work in pairs to let your carefully control the movement of the bones. This picture shows that you have two muscles to control movement of your knee.

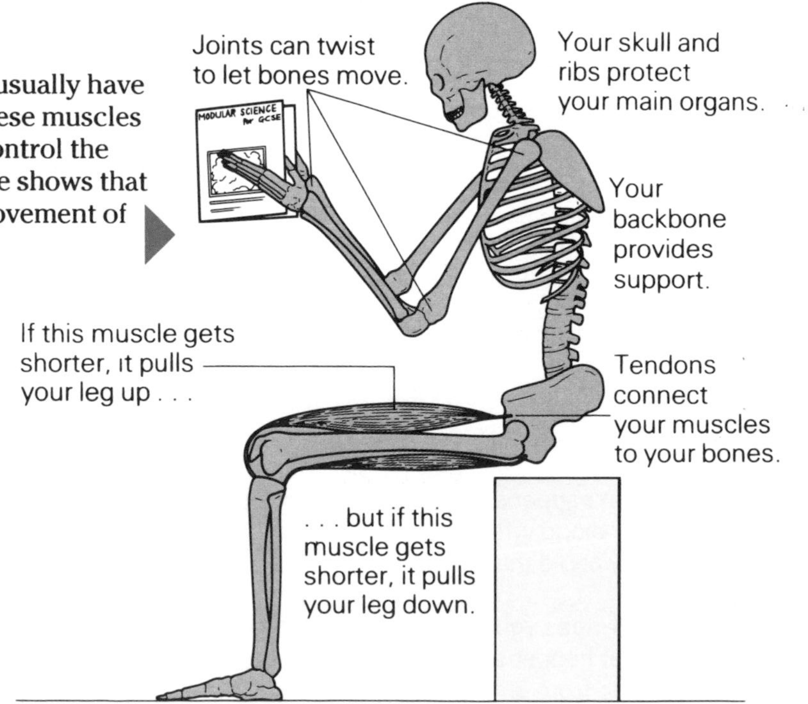

If you want to keep your leg halfway up, both muscles will have to do some pulling. The top muscle will have to pull to keep your leg up. But the bottom muscle will have to pull to stop your leg from going up too far. The two muscles will have to work together, but pull *against* each other. When this happens, they are called an **antagonistic pair** of muscles.

Under careful control

How many ways can you bend your knee? The knee is a fairly simple joint – it can only twist forwards and backwards. What about your shoulder, wrist and neck? These joints are more versatile – they can move in many directions.

The bones which make up these joints have many antagonistic pairs of muscles attached to them. This allows you to control the movement of your arms, hands and head with accuracy. Imagine how difficult it would be to write a letter with a pen tied to your knee!

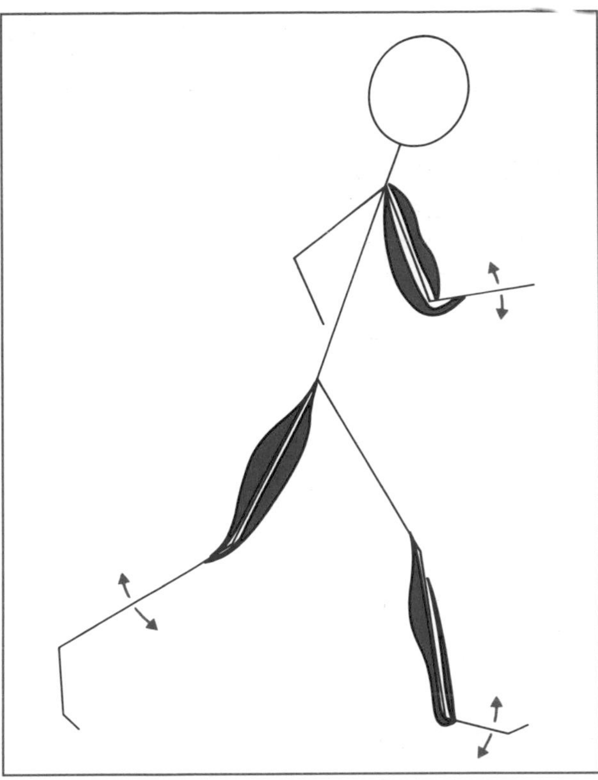

There are many antagonistic pairs of muscles in your body – and they often work in groups.

Taking the strain

When you hurt some part of your body, you try not to use it. For example, if you hurt your foot, you tend to limp, keeping your foot stiff to avoid bending it as you walk. This often makes other muscles in your leg and back start to feel sore. This is because these other muscles are part of a group of antagonistic pairs of muscles used in walking. By limping, you keep the strain off of your foot but you put extra strain on to other groups of muscles.

'Joint Action'

It is impossible for a single joint to be able to move in every direction – if it tried to, the bones would be twisted completely off the joint. But you can move your hand to touch anything around you. How do you manage to do this?

Think for a moment of how you get through a crowd of people. You can't usually walk in a straight line. You have to turn left and right as you walk along – but you get through. The same is true of moving your hand to a particular place – you may have to twist your shoulder a bit, bend your elbow and move your wrist. By moving more than one joint at a time, almost any movement is possible.

Like most movements, a gymnast's routine uses several joints at a time.

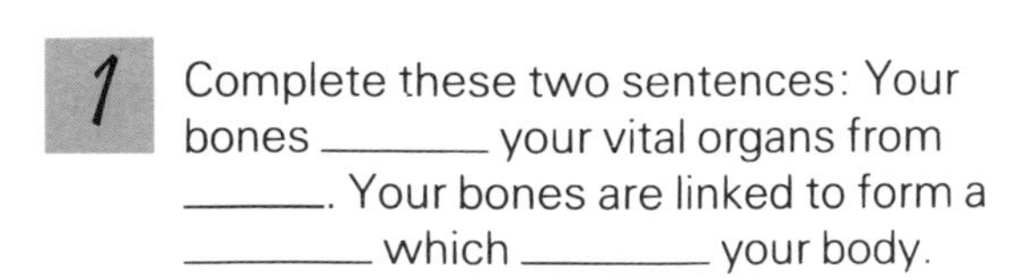

1 Complete these two sentences: Your bones ________ your vital organs from ________. Your bones are linked to form a ________ which ________ your body.

2 What causes your muscles to change length?

3 When you clean a blackboard, which joint twists the most? Draw a diagram to show how an antagonistic pair of muscles could make your arm move in this way.

4 Imagine your right arm is stretched out sideways. Which joints will you have to use to move your right hand *over* your head to scratch your left ear?

3.11 Getting hurt

Making a wrong move

Your body is always moving – even when you are asleep. These movements may be strong and fast, especially when playing sport. If your bones and joints are in the wrong position when you move them, you can get injured. The most common serious injuries are sprains, dislocations and fractures. If they are not treated immediately, these injuries can get much worse. People who are trained in 'First-Aid' usually know how to treat such injuries.

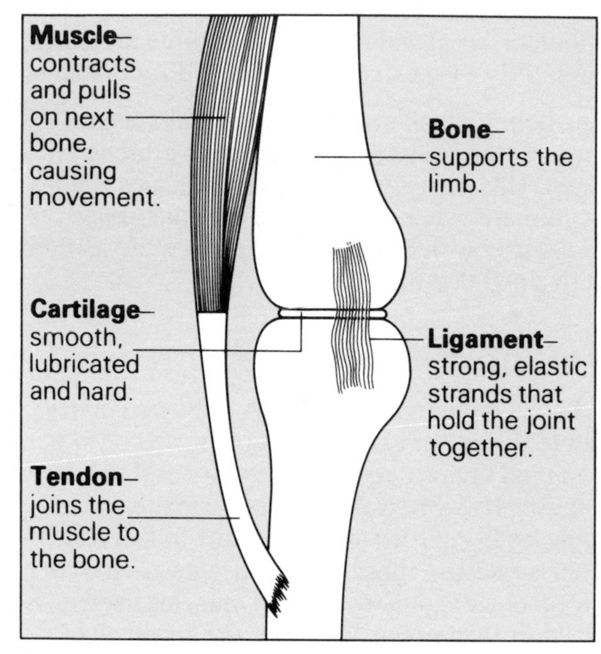

A typical joint – without your joints, you would not be able to move at all.

Around the bend

Your joints are made up of several parts. A joint connects two bones; these bones are usually separated by a piece of tough **cartilage**. The bones are held together in the joint by strong elastic strands called **ligaments**. Muscles make the bones move around the joint. These muscles are fixed to the bones by **tendons**. The tendons are usually found very close to the joint.

A hockey injury

The player's knee is damaged. What could have happened?

Forcing a reaction

If a strong sideways force acts on a joint, you may **sprain** the joint. This happens when the ligaments and tendons are over-stretched. The joint usually reacts by swelling up and becoming stiff and sore. By making it difficult or painful to use, your body protects the sprained joint. This is because it makes you rest the injury.

Sprain

Knee cap

Ligaments and muscles across the joint are over-stretched

Dislocation

Bones no longer meet at the joint.

Lower leg bones

Upper leg bone

All out of joint

If the sideways force is very strong, the bones may be pulled completely out of place. This is called a **dislocation** – this is because the bone has been separated from the joint. All ligaments and tendons will be greatly over-stretched. It is important that only a trained person treats a dislocated joint – and quickly. This is because any further movement could easily tear the ligaments and tendons.

Sticks and stones....

Very large forces will **fracture** your bones. A hard fall or a fast moving hockey stick or football boot may be enough to do this. Even the force from your own muscles can fracture your own bones. Some unlucky footballers have broken their kneecaps by kicking the ground instead of the ball. The muscle which makes the leg kick is attached to the knee cap. By kicking the ground, the leg stops moving – but the muscle keeps pulling and breaks the kneecap.

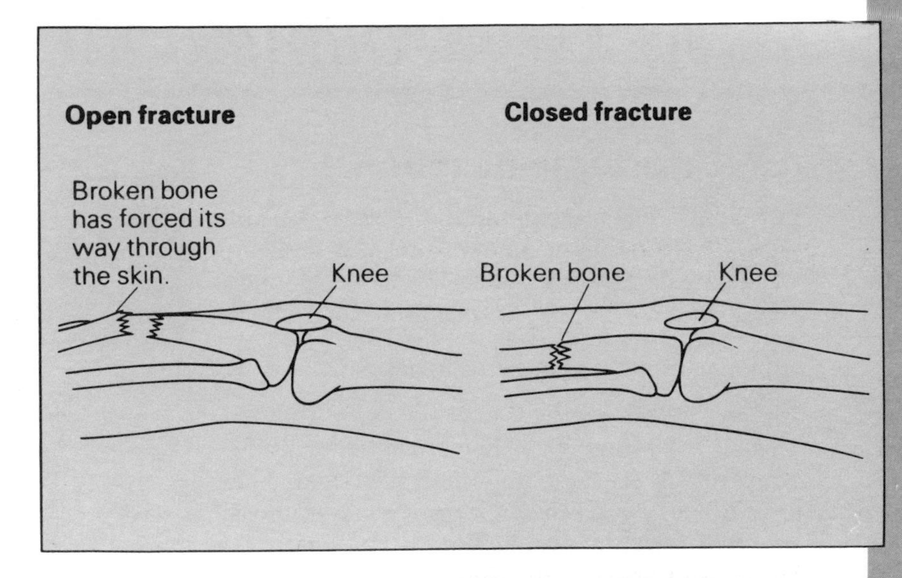

Help at hand – First Aid

First-Aid consists of various simple, sensible things you can do when someone is injured. It usually involves stopping injuries from getting worse. Serious injuries may need treatment by a doctor, but First Aid can help long before the doctor gets a chance to treat the injury.

Wherever there are a lot of people, someone may get injured. At sports events and pop concerts, St. John ambulance staff provide First-Aid. In factories and large offices, some of the workers will be trained as 'First-Aiders' – so will some of the teachers in your school.

Simple First-Aid is easy to learn, but serious injuries need trained help.

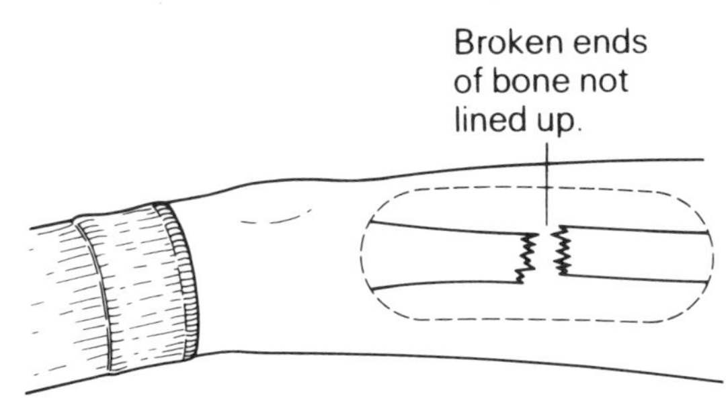

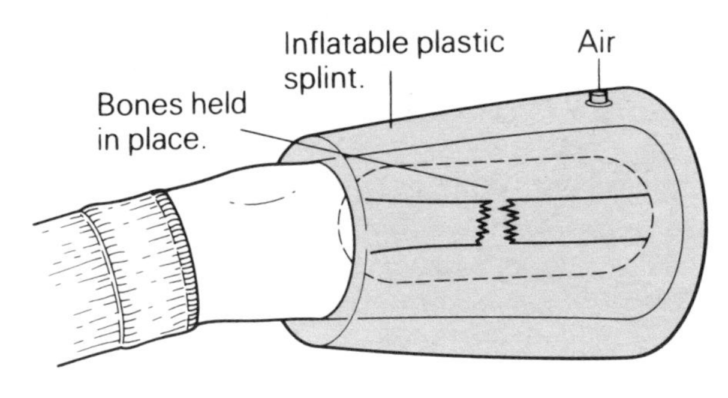

Any movement may cause the jagged broken bone to cut arteries, nerves or flesh. The splint supports the broken bone and lines the ends up preventing further damage.

1. Your bones and joints can suffer three types of serious injury. What are they?
2. Which five things are connected together to make a joint?
3. How does your body try to protect damaged joints?
4. What is the difference between a closed fracture and an open fracture?
5. Find out if any of your family or friends have ever suffered from a sprain, dislocation or a fracture? Try to get information about at least ten people's injuries. Draw up a table to show their age at the time of the injury suffered and how it was caused. Which is the most common injury?

3.12 Controlling your reactions

An Automatic Reaction

Your body has to defend itself from attack all the time and sometimes it has to react very quickly. When you blink suddenly, you may be defending yourself against something moving towards your eye. You do not have time to think about this blinking action, it just seems to happen 'automatically'!

This type of blinking can happen in one tenth of a second – this is very quick, but the blinking has to be quick. If you blinked more slowly, your eye would be hurt more often. Your sight is very important to you, so it is very difficult to stop this reaction.

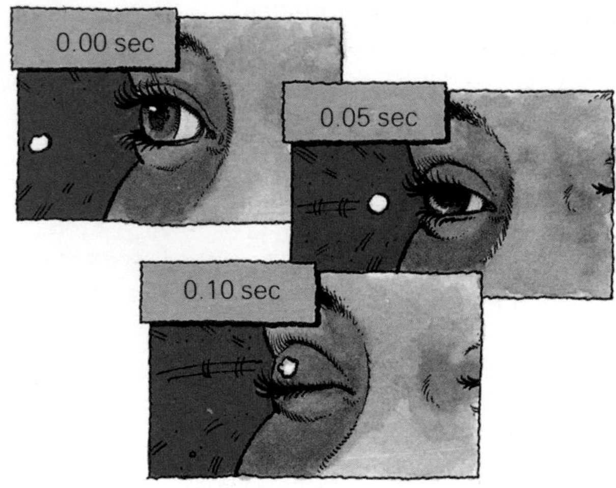

Even if you didn't notice something thrown at your eye until very late, your eyelid could still close in time.

Short circuiting the system

How can your senses get your muscles to react so quickly? Your body contains millions of **nerves**. These can receive and send little electrical messages to and from each other. Each of these nerves has a special job to do.

Some nerves can detect changes in the surroundings. These are called **sensory nerves**. These usually send messages all the time way back to your brain. This lets you decide what to do with the information from the nerve.

But sometimes the message does not reach your brain. If the message needs urgent action, it can get short-circuited by a **relay nerve**. These are found in your spine. The relay nerve sends the message straight to a nerve which controls a muscle. This type of nerve is called a **motor nerve**. When a motor nerve gets a message, it makes the muscle move.

The message makes a 'short-circuit' by going through the relay nerve instead of going through your brain. This 'short-circuit' means that the muscle can move much more quickly. This is how your body can react 'automatically' to sudden changes in your surroundings.

Your reflex reactions are much faster than reactions you have to think about – you will drop a hot tray before you can shout in pain.

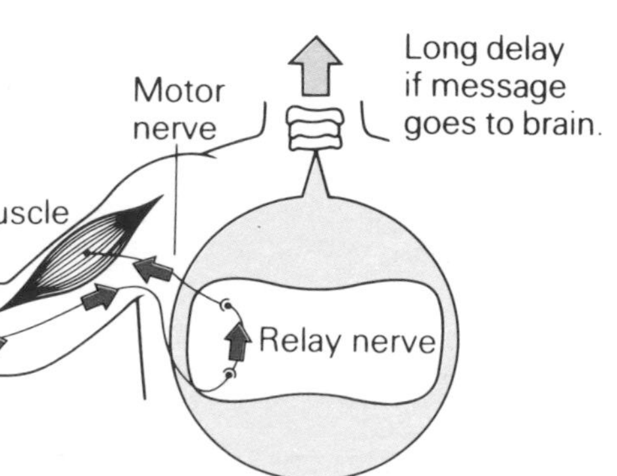

Relay nerves allow you to have very quick reactions.

Sensible Precautions

Your body has several 'warning systems' to protect you from injury. Your eyes are one example – how do they help to protect you? There are five such warning systems – sight, hearing, touch, taste and smell. These are called the five **senses**.

These senses provide you with information about your surroundings. You then use this information to guide your actions. For example, rotten food can make you sick, but if you start to eat some rotten food, you can quickly taste the rotten flavour. Your sense of taste has been able to warn you not to eat the food.

Your sight tends to dominate the other senses, but it can't do everything. Although you can only look in *one* direction at a time, your hearing 'listens' in *all* directions. You may not be able to see if something is hot or sharp, but a gentle touch will soon tell you.

Her sight shows her where to reach for the bar, so her eyes help to 'protect' her from falling.

Making sense of your senses

Very young children have to learn how to use their senses to control their actions. They are often slow or clumsy because they cannot make good use of reflex reactions. For example, they fall over a lot because they cannot control their muscles quickly enough to stay balanced upright. When they are older, they will run or walk or hop without any problems – because they will be using reflex reactions to keep their balance without even thinking about it.

Regular practice can develop new reflex reactions and can improve the speed of existing reflexes. Most sports require quick reactions of some kind – and the more you practice, the better you get at the game. This is because practice develops the particular reflex reactions which you need to be able to do well at the sport.

He won't catch the balloon because he can't control his eye reflexes – his eyes close when anything is near his face.

1
a What are the five senses?
b Which parts of your body are involved with which senses?

2
a Explain what sensory nerves do.
b Explain what motor nerves do.

3 How can a sense of smell help to protect to you?

4 Give examples of at least five kinds of reflex reactions. In each case explain what would happen if there was no reflex reaction.

5 Find out two ways in which blind people overcome their lack of sight. Which senses do they use to help them do this?

3.13 Measuring fitness

These people are all very fit, but they would soon get tired doing a different sport.

Fit for what?

There are many different ways of describing how fit you are. You could be fit in terms of speed or stamina or agility. A sprinter may be able to run very fast over a short distance but may get very tired if he or she has to run several miles. A marathon runner can run for hours but may get exhausted by a little weight-training. So what does being 'fit' mean?

During any activity you do, your heart will be beating and you will be breathing. This means two good tests for fitness are:

- How fast does your heart go when you are active?
- How fast do you breathe in and out when you are active?

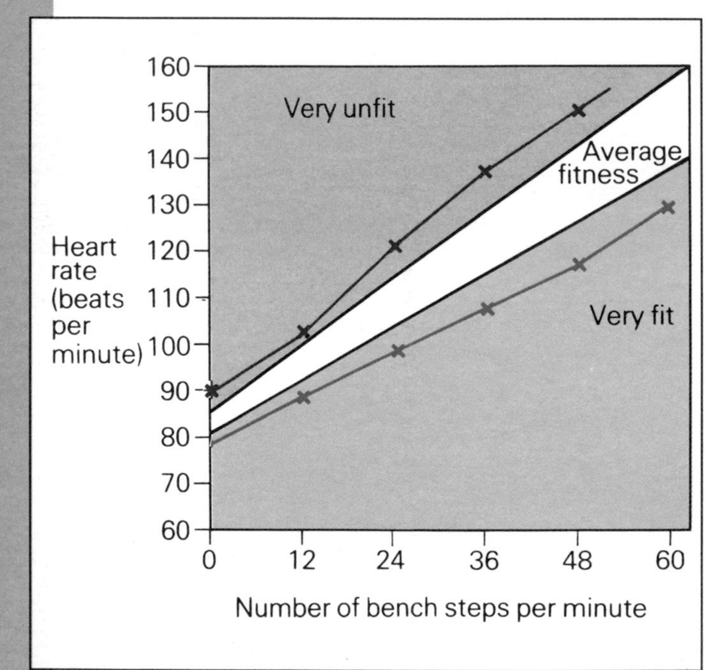

Heart rate increases when you are active. The red line gives the results for an unfit boy (weighing 50kg). The blue line is for a fit boy of the same weight.

Activity and your heart

When you are active, your body has to work harder than usual. You have to use up a lot of energy to make yourself move – this energy comes from respiration. (You can read more about respiration on page 72.)

You need oxygen and glucose to help you to release energy during respiration. The energy is needed by your muscles, to make them move. How can the oxygen and glucose get to your muscles?

The oxygen and glucose are carried around your body by your blood. The more active you are, the more energy you need. This means you need to supply more oxygen and glucose to your muscles. To do this, your heart must pump more blood around your body. So, the more active you are, the faster and harder your heart will beat. A fit person's heart can do this easily – but an unfit person will put a great strain on their heart.

Activity and your lungs

How do you get a lot of oxygen into your blood when you are active? You need to breathe in and out more quickly *and* more deeply. Deep breaths in and out mean you take in more fresh air and get rid of more used air.

Breathing out is very important – it gets rid of waste carbon dioxide from your blood. The more energy you use up when you are active, the more waste carbon dioxide is produced. This has to be got rid of. If you don't get rid of it, you will feel dizzy and faint.

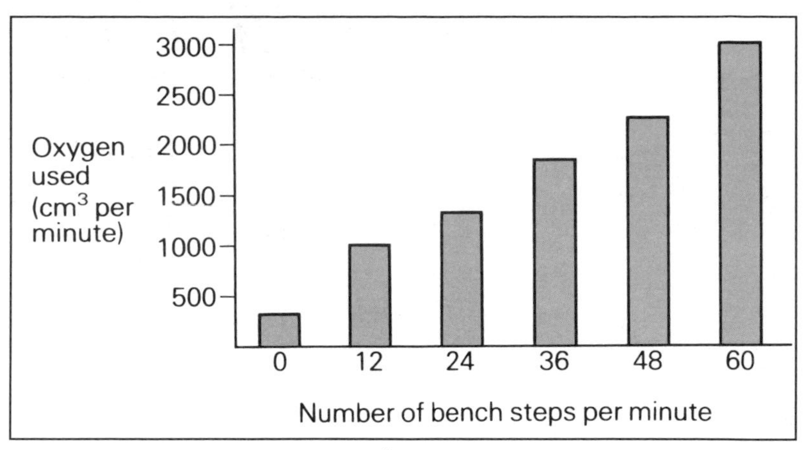

When resting you use about 300cm³ of oxygen a minute. How many times more oxygen do you use after doing 60 bench steps in one minute?

How fit are you?

You can test your fitness by finding out exactly how much your heart rate increases when you become active. If you are fit, your heart rate will be low to start off with. As you exercise, your heart rate will not increase by much. You can check this on the heart rate graph on the previous page.

Another way of checking your fitness is to find out your **recovery time**. This is the amount of time that you have to wait for your heart rate to return to normal. Your recovery time gets shorter as you get fitter.

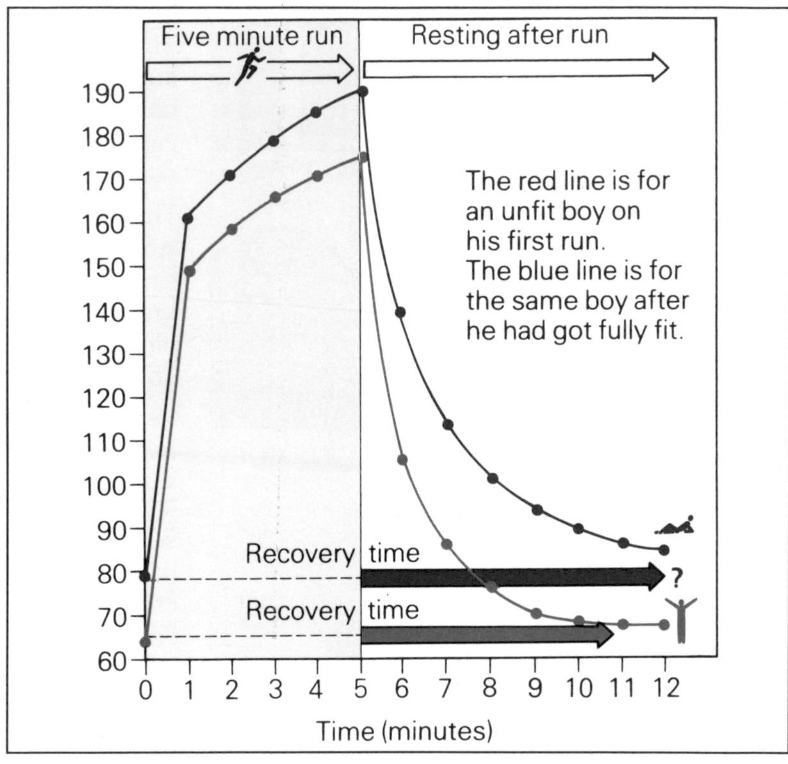

By measuring heart rate during and *after a five-minute run you can find out your recovery time.*

1 What are the two best ways to test for fitness?

2 Why does your blood need to move more quickly when you are active?

3 Why do you produce more carbon dioxide when you are active?

4 **a** A boy has a heart rate of 130 beats per minute after doing 48 bench steps in 60 seconds – is he fit or unfit?
b If his heart rate was 160 beats a minute would he be fit or unfit?

5 Think of a simple (*but safe*) way in which you could measure the volume of air you breathe out. Describe your idea. Use a diagram to show how it could be done.

3.14 Keeping fit

It's up to you

If you take care of your body, you should remain fit and healthy. But if you neglect your health, your body can be seriously damaged within just a few years.

In Europe and North America, disease and malnutrition are not common – thanks to modern medicines, clean water supplies and a good supply of food. But serious health problems still occur. Why should this be so?

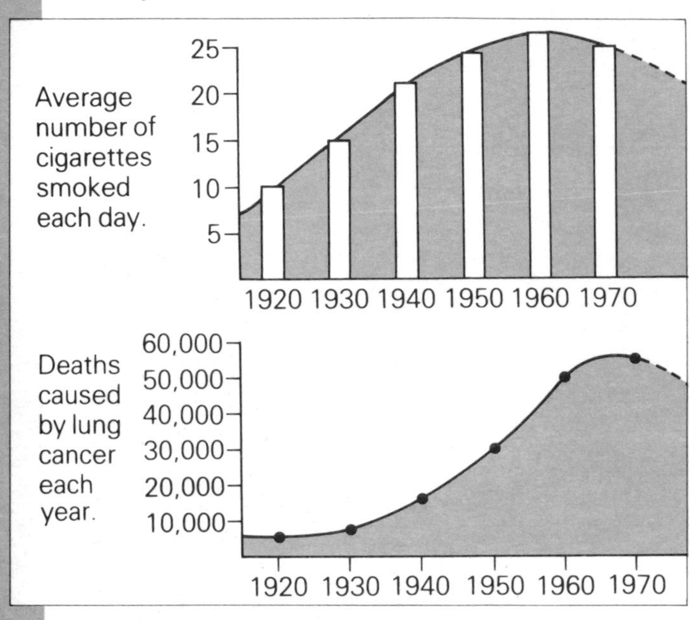

Increases in cigarette smoking and increases in lung cancer over the same period.

Smoking

When someone smokes a cigarette, their heart rate gets faster. Their blood vessels get narrower and this makes the flow of blood more difficult. The increase in heart rate and the poor blood flow means that their heart has to work much harder to pump blood around the body. Also carbon monoxide gas in the smoke poisons some of the muscles of your heart.

The cigarette smoke contains tar and large smoke particles. These get stuck in the small air sacs in your lungs. This makes it difficult to breathe easily. Smoking weakens your heart and blocks up your lungs – so if you want to be fit, don't smoke!

As well as affecting your fitness, smoking is a major cause of several lung diseases. These diseases, such as lung cancer, will eventually kill you.

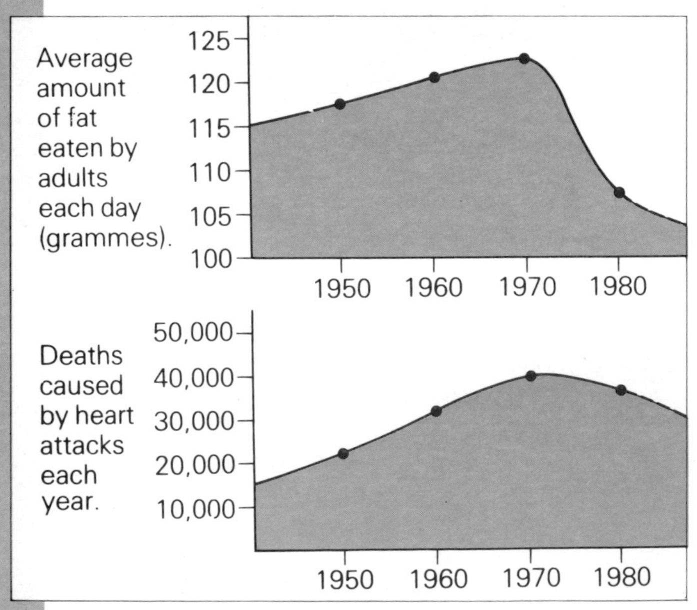

After many adults reduced the fat in their diet, the number of deaths due to heart attacks started to fall.

Diet

Some of the fats produced from your food can stick to the walls of your arteries. These fatty deposits can get larger and may even block an artery. The coronary artery supplies blood to your heart muscles. If this artery becomes blocked, your heart will stop beating. This is called a heart attack and can be fatal.

Not all fats cause these problems. There is a special group of fats called **polyunsaturated fats**. These are safe to eat because they do not block arteries.

You should exercise regularly to keep the heart muscles in good condition. Any exercise will do – as long as it makes your heart rate go up. A lack of exercise means that the heart does not get strengthened. Your body will also become fat and heavy. This extra weight will put a greater strain on your body, and also your heart.

Alcohol

Alcohol is a drug which affects your brain and your reflexes. This is why people who are drunk slur their words and become clumsy. The capillaries below the skin become larger and make the skin look red. In large or frequent doses, alcohol is a poison which can seriously damage your liver and nerves. But small doses make people feel relaxed and this is why it is a widely used drug.

Many people drink alcohol in pubs, but drivers should not drink and drive.

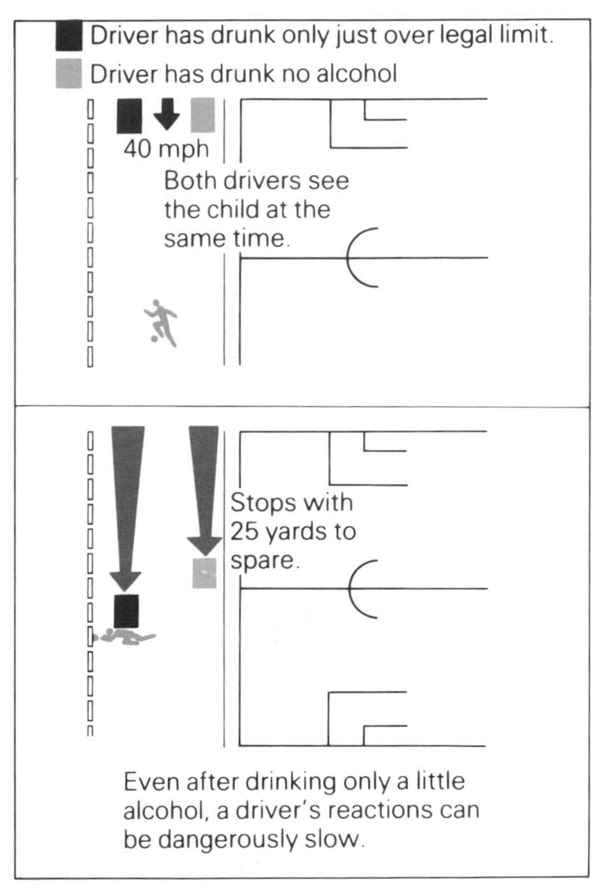

Alcohol slows the reflexes and drivers cannot stop quickly enough in an emergency.

This man is blowing into an electronic breathalyser. The police officer will now be able to tell if he is too drunk to drive.

1 Write down four ways in which smoking makes your heart do extra work.

2 What has happened to
- **a** the amount of tobacco smoked each year since 1920?
- **b** The number of deaths due to lung cancer over the same time?
- **c** Do the graphs suggest any link between smoking and lung cancer? Explain your answer.

3 Look at the graphs which show fat intake and heart attacks. Do you think the two may be related? Explain your answer.

4
- **a** Why is alcohol a popular drug?
- **b** Why is it dangerous?

5 Look at the braking diagram above. What does it tell you about the risks of 'drinking and driving'?

3.15 What a life you lead!

Decisions, Decisions

You begin making decisions about your life at a very early age. For example, whether to eat your dinner or to go out to see your friends. Even these decisions can affect your life. Having 'no dinner' may mean you go hungry for the rest of the day. There are many people in the world who cannot make decisions about the food they eat. They have no choice – because they can't get any food. Forty million of them die each year from starvation. That's equal to two out of every three people in Britain.

This child may be lucky to get a decent meal soon. How many times in a week are you that lucky?

Decisions and consequences

You may be fortunate enough to be able to make decisions about your life. You may also be unfortunate to suffer the consequences. The case studies show how three people live their lives. They have to make decisions about their lives all the time...just like you.

Case studies

The following descriptions about Jamie, Meta and Simon are from other members of their class, read the descriptions carefully and pick out as much information as possible.

Jamie is 14 and plays football for the school and the Town team. He trains about 4 times a week and usually plays twice a week.

His parents make sure he eats the right foods and even give him a special packed lunch. Although he plays a lot of football, he doesn't ever seem to get tired out.

He is friendly with some older boys and has started drinking cider and smoking cigarettes on his way home from training and matches.

Meta is 16 and is usually top of the class. She does not join in the sports activities because she is often too tired.

She hardly eats at all during the day and her best friend says that after meals at home she is sick.

She says she is cold and she always looks ill.

She is very fussy about her food and doesn't like to eat fruit and vegetables.

Simon is nearly 15 and wishes he wasn't at school. He is overweight and always tries to get out of sports.

He likes to **watch** sports but hates to play them. He usually watches us from the touchline, eating sweets and chocolates while we run around.

If he has to run for a bus, he is always the last there and he gets really hot and goes red in the face.

What would you advise Jamie, Meta and Simon to do so as to live a healthier life? Make a life plan for one or more of the people. In each life plan:

a) list the things they do that **may do them harm;**

b) list the things that are **good for them;**

c) plan what they should do to have a **life they will enjoy.**

You may find some useful ideas by looking back through this module.

Index

A

acid rain **36–37**
acid strength **36**
activity (effects of) 86–97
air 16
air pollution 17, **32–33**, 34–39
air sacs **73**, 88
alcohol 89
alkali **37**
aluminium recycling 89
ammonia **78**
antagonistic pairs (muscles) **80**, 81
aquarium 18
arsenic (pollutant) 43
arteries **76**, 77, 78

B

background radiation **48**
bacteria **13**, 16, 18, 40, 41
balanced diet **68**
biodegradable plastic **59**
birds (feeding habits) 7, 9
birds and pollution **44**
blood **74–75**
blood transfusion 74
blood groups 74
blood vessels 70, 73, **76–77**, 78
body temperature **64**
bones 80, 82, 83
bottle-bank 58
breathing 16, **72–73**, 87
breathalyser 89
broken bones 83
bronchitis 35, 52
burning 16, 17

C

cadmium (pollutant) 35, 46, 47
cancer 46, 47, 48, 49, 52, 62, 88
capillaries 77, 89
carbohydrate **11**, **66**, 68 (*see also* starch, sugar)
carbon 16
carbon cycle **17**
carbon dioxide **10**, 11, 16, **17**, 19, 34, 36, 72, 73, 76, 78, 87
carbon monoxide 34, 75, 88
carcinogens 47
carnivores **5**, 6, 7, 25
cartilage 82
catching animals 20
cells 63, 65
cellulose **11**
cement dust 43, 46
Chernobyl 48
chlorophyll **10**, 11
cigarettes **52–53**, 88
coal 16, 33, 46
colonisers **9**
combustion **16**, 17
community **5**, 7, 18, 20
competition **9**, 23, 25, 51
compost **15**
consumers **5**
control (of reflexes) 83, 84
converters (car exhausts) 37
core temperature **64**
coronary arteries **88**
cumulative poison 24–25, 26–27, **35**, 50

D

decibels **53**
decomposers **5**, **14–15**, 16, 18, 41
diaphragm **72**
diet 7, **68**, 88
digestion **71**
disease 41
dislocation (joint) **82**
drip feed 70
dumping 49, 57
dust (pollutant) 46, 47

E

enzymes **70**, 71
energy (from sun) 10, 12
(from sugars, glucose) 11, 19, 72, 86
(from food) 12, **64**, 66, 68
energy flow 13
energy loss 13, 64, 79
energy use 12, 64, 65
environments **2**–31, 32–33
exercise 86, 87, 88
excretion **78–79**

F

farming, farmland 22–23, 24, 26
fats **66**, 68, 77, 78
fertilisers 15, 43
fibre **67**, 71
first-aid 83
fitness 86–87
fossil fuels **16**, 17
food **66–67**
food chain **6**, 12, 13, 25
food web **7**, 18
fracture (bone) 83
fungi **13**, 14, 16, 18, 51, 59
fungicides **51**

G

gas (natural) 16, 35
glucose **70**, 72, 86
growing 12, 62, 65, 66

H

habitats **4**, 9, 22, 23
haemogloblin 66, 67, 73, 75
health 35
heart **76**, 86, 87
heart attack 88
heart muscles and smoking 53, 75, 88
hedgerows 2, **22–23**
herbicides 51
herbivores **5**, 6, 7
hypothermia **64**

I

incineration **56**, 57
indicators (chemical) **19**
(plant) **38**–39
(animal) **42**
injuries **82–83**
insecticides 3, **24**–25, 51
intestines (small) **71**
iodine (mineral) 67
iron (mineral) 66, 67

J

joints **80–81**, 82

K

kidneys 78, **79**

L

land filling **56**, 57
land use **22**
lead (pollutant) 34
lead-free petrol 35

leaf 10, 11
leukaemia **48**, 49
lichens **38**–39
life (features of) **62**
(energy for) **65**
ligaments **82**
lime 37
Lincoln index 21
liver **78**
lung cancer 46, 49, 52, 88
lungs 16, **72–73**, 76

M

malnutrition **69**
manure 15
mercury (pollutant) 35, 46, 47
minerals **66**
motor nerve **84**
mouth 71
movement 12, 13, 65, 80, 81
muscles **80**, 81

N

natural radiation **48**
nerves **84**
nickel (pollutant) 43, 46, 47
nicotine 52, **53**
nitrates 15, 18, 43
nitrogen dioxide (pollutant) 34, 36, 37
noise **54–55**
nuclear power, nuclear waste **48–49**

O

oil 16, 43, 44–45
omnivore 5
organism **62**
otter **24–25**
oxygen 10, 11, 16, 19, 41, 72, 73, 75, 76, 86, 87

P

pH **36**
paper factories 42
passive smoker 53
pest control 51
pesticides 50–**51**
petrol 34, 35, 43
phosphates 15, 43
photosynthesis 6, **10–11**, 13, 16, 17, 18, 19
pitfall trap 20, 21
plasma (blood) **74**
pollution **32–61** (*see also* air-, dust-, noise-, nuclear-, pesticide-, sea-, water-pollution)
pollution indicators 38–39, 42
polyunsaturated fats **88**
pondweed 19
populations **20**, 24–25, 26–27, 50
power stations 33, 46
producers **5**, 6, 10, 18
proteins **66**, 68, 78

Q

quadrat **20**

R

radiation **48**
radioactive waste **48**, 49
reactions (nervous) **84**, 89
recovery time **87**
rectum 71
recycling **15**, 17, 18, **58–59**
red blood cells **74**
reflex 84
relay nerves 84
respiration **11**, 13, 16, 17, 19, **72**, 76, 86
rodenticides **51**
rubbish 32, 56–57, 58–59

S

sampling a habitat **20**, 21
sand dunes 4, 28
scrubbing gases 37
sea pollution **44–45**
senses 85
sensory nerves 84
sewage **40**, 41, 42
shunt vessels 79 (*see also* alcohol)
skeleton **80**
skin 78, **79**
small intestines **71**
smoke 34
smokeless fuel 35
smoking (cigarettes) 52, 88
soap 43
soil nutrients 9, 11, 15, 23
soot 34
sound 54
sound-proofing 55
sprain **82**
starch **11**, 70
stomach 71
sugars (in photosynthesis) **10–11**
sulphur dioxide (pollutant) 34, 35, 36, 38, 39, 46)
sunlight (energy from) 5, **10**
swans 26–27
sweat glands **79**
sweep net 20

T

tar (cigarette) 52
tendons 82
tobacco 52

U

urea **78**, 79
urine 79

V

valves (heart) 76, 77
veins 76, 77
vitamins **66**

W

warmth (body) 12, 13, 64
waste (natural) 13, **14–15**, 17, 18, 32
water (in photosynthesis) **10–11**
(in respiration) **11**, **72**
water pollution 40–41, 42–43
Water Authorities 25, 43
weeds 8, 9
white blood cells 74
woodlands **4–5**, 6, 9

Acknowledgements

The authors and publishers would like to thank the following people for their contributions to the development of the **Modular Science for GCSE** series:

Martin Stirrup, Chris Kitchen

The extract from 'Tarka the Otter' on page 24 is reproduced by the kind permission of Bodley Head Ltd.

Photo Acknowledgements
Altrincham Council *56 (centre);* Barnaby's *64 (bottom), 68 (top, bottom), 89 (top)*; Biophoto Associates *5 (bottom)*; J. Allan Cash *33 (right), 46, 54, 61 (bottom), 62 (top, centre), 68 (centre)*; Bruce Coleman *3 (centre), 17, 18, 28 (bottom)*; Friends of the Earth *59 (bottom)*; Geo Scientific Films *9, 14 (top, bottom right), 16, 19, 23, 27 (top), 28 (top left), 34, 44 (bottom), 52*; Sally & Richard Greenhill *53 (bottom), 55 (top)*; Greenpeace *32 (left, bottom right), 37 (top), 49*; Health Education Authority *52, 63, 63*; John Hillelson Agency *47*; Holt Studios *15, 33 (bottom), 43, 51 (top, centre), 55 (bottom)*; E. & D. Hosking *5 (top), 11, 14 (left), 25 (bottom), 32 (centre), 44 (top), 50, 56 (centre), 61 (top)*; Hutchinson *90*; Impact Photos *36, 36*; I.C.C.E. *3 (top), 4, 25 (centre), 28 (top right), 31, 37 (centre), 57, 58, 59 (top)*; Carmilla Jessel *84*; Frank Lane Agency *Cover, 2, 3 (top left, bottom), 14 (centre), 22, 24, 26, 27 (bottom), 29, 32 (top), 51 (bottom)*; Metropolitan Police *89 (bottom)*; Panos Pictures *41, 42*; Chris Ridgers *61 (right)*; St John's Ambulance *83*; Science Photo Library *33 (centre), 38, 38, 38, 53 (top, top), 62 (bottom), 65, 65, 65, 65, 65, 70, 74, 75*; Sporting Pictures *64 (top), 81, 85, 87, 87, 87*; Peter Vinc *80*;

Covers: B Withers. *Title page:* Geo Scientific Films; Science Photo Library; Sporting Pictures. *Contents pages:* Geo Scientific Films; E & D Hosking; Science Photo Library.
Agency Photographers: *Greenpeace* - Kavert, Zindler (32), Venneman (37), Gleizes (49); *ICCE* - P Steele (3, 4, 28, 31, 59), M Boulton (25, 57), C Agen (37), M Hoggat (58); *Frankie Lane Agency* - B Withers (Cover, 29), J Watkins (2), M Thomas (3, 22), L Batten (3), L West (14), A Hamblin (24), J Lynch (32), S Malowski (51); *Panos Pictures* - G Castro.
Picture Research - Jennifer Johnson

Heinemann Educational Books Ltd.
Halley Court, Jordan Hill, Oxford OX2 8EJ

OXFORD LONDON EDINBURGH
MELBOURNE SYDNEY AUCKLAND
SINGAPORE MADRID ATHENS
IBADAN NAIROBI GABORONE HARARE
KINGSTON PORTSMOUTH (NH)

ISBN 0 435 57000 5

First published 1988

Designed and phototypeset by Gecko Limited, Bicester, Oxon

Printed in Great Britain by Scotprint Ltd, Edinburgh